THE

STRUCTURE OF ANIMAL LIFE.

SIX LECTURES

DELIVERED AT THE BROOKLYN ACADEMY OF MUSIC IN
JANUARY AND FEBRUARY, 1862.

By LOUIS AGASSIZ,

PROFESSOR OF ZOÖLOGY AND GEOLOGY IN THE LAWRENCE SCIENTIFIC
SCHOOL.

Louis Agassiz

Louis Agassiz was born on 28th May 1807 in Môtier (now part of Haut-Vully), in the canton of Fribourg, Switzerland. He is best known as a celebrated biologist, geologist, physician and controversial racial theorist. Educated first at home, then spending four years of secondary school in Bienne, he completed his elementary studies in Lausanne (a city in Romandy, the French speaking part of Switzerland). Having adopted medicine as his profession, Agassiz studied successively at the universities of Zürich, Heidelberg and Munich; while there he extended his knowledge of natural history, especially of botany. In 1829 he received a PhD at Erlangen, and in 1830 that of doctor of medicine at Munich.

In 1832 Agassiz was appointed professor of natural history in the University of Neuchâtel (near Berne). The fossil fish there soon attracted his attention, and as early as 1829, Agassiz planned the publication of the work which, more than any other, laid the foundation of his worldwide fame. Five volumes of his *Recherches sur les poissons fossiles* ('Research on Fossil Fish') appeared at intervals from 1833 to 1843. They were magnificently illustrated, chiefly by Joseph Dinkel. Agassiz found that his palaeontological labours made necessary a new basis of ichthyological (the study of fish) classification. The fossils rarely exhibited any traces of the soft tissues of fish, consisting chiefly of teeth, scales and fins, sometimes even bones. He therefore adopted a classification which divided fish into four groups: Ganoids, Placoids, Cycloids and Ctenoids, based on the nature of the scales and other dermal appendages.

While Agassiz did much to place the study of fish on a rigorously scientific basis, this classification has been superseded by later work. In other achievements, in 1837, Agassiz was the first to scientifically propose that the Earth had been subject to a past ice age. With the publication of his *Etudes sur les Glaciers* ('Study on Glaciers') in 1840, he discussed the movements of the glaciers, their moraines, influence in grooving and rounding the rocks over which they travelled, and in producing the striations and *roches moutonnees* seen in Alpine-style landscapes. This work has since been highly influential. Two years later (continuing until 1846), Agassiz issued his *Nomenclator Zoologicus*, a classified list, with references, of all names employed in zoology for genera and groups — a work of great labour and research.

After these successes, Agassiz travelled to America in the autumn of 1846, with the twin purposes of investigating the natural history and geology of North America and delivering a course of twelve lectures on 'The Plan of Creation as shown in the Animal Kingdom' at the Lowell Institute in Boston, Massachusetts. Agazzis settled in America, continuing lectures for the Lowell Institute, and was appointed professor of zoology and geology at Harvard University in 1859. After coming to the United States, he started writing on the genre of 'scientific racism', advocating 'polygenism', a belief that races came from separate origins (specifically separate creations), endowed with unequal attributes, and could be classified into specific climatic zones, in the same way he felt other animals and plants could be classified.

In recent years, critics have cited Agassiz's racial theories, arguing that these now-disproved and unpopular views tarnish his scientific record. This has occasionally prompted the renaming of landmarks, schoolhouses, and other institutions which bear the name of Agassiz (which abound in Massachusetts). Stricken by ill health in the 1860s, he resolved to return to the field for relaxation and to resume his studies of Brazilian fish. In April 1865 he led a party to Brazil and in December 1871 made a second eight month excursion, known as the *Hassler* expedition, visiting South America on its southern Atlantic and Pacific seaboards. This scientific voyage drew the praise of Charles Darwin, whose evolutionary theories Agassiz vehemently rejected throughout his life. Agassiz died on 14th December 1873, aged sixty-six in Cambridge, Massachusetts, America.

PREFACE.

In bequeathing to the Brooklyn Institute a fund the interest of which was to be applied to the purchase of pictures by the best American artists, Mr. Graham, no doubt, proposed to himself gradually to build up a gallery in Brooklyn which should show to future generations the state of the arts in the past. So also, in these lectures, which we have denominated the "Graham Lectures," he designed to supply a series of lectures by men of the highest literature and science, which, by being published, should furnish a series of authors that would evince to future ages the progress of thought and talent among their ancestors, and become a valuable part of the literature of our country.

For these lectures he designed that the proofs of the "Power, Wisdom, and Goodness of God" should be drawn directly from His works, as they exist in, and in fact constitute nature. The obvious teaching of the creation can hardly be questioned. Every man is, in some measure, an admirer of the works of God. From the moment when, as a child, the flowers attract his attention, to mature years, Nature is presented to him, in her various forms, harmonies, and adaptations, and becomes a subject of study and admiration, "especially to one who," as Saint Pierre remarks, "stops at every step, transported by the beauty and sublimity of her divine works."

Students and men of science acquire a more minute knowledge, and learn those interior facts which escape the common observer. The man of science, therefore, will trace the hand of God, His skill, goodness, and wisdom, at every step of progress in his studies. He will discover more minutely that sublime wisdom which has made the wonderful organisms of plants and animals, and arranged the immensity of nature.

Modern science has very much extended the fields of study by its discoveries, and has greatly enlarged our knowledge of the extent and variety of creation.

Not only does the profound thinker observe the harmonies and adaptations of God's works in the visible creation, but he is led to study the no less wonderful harmonies to be found in the intellect, the morals, and the affections of men. All become to him the subjects of study and admiration. Nature, in the largest sense of the word, may be considered to include all created things, with their habits, and the laws by which they are governed, and by which they operate. Thus all that we call art is but the effect of the natural laws which the Creator has given to his creature man, and which He has stamped upon his nature. Thus the works of man, as well as those works with which he has no power to interfere, are but a part of God's design. The works which grow out of man's skill are derived from the nature and capacity bestowed upon man by his Creator. The artistic dwelling, and the steamship, no less than the house and dam of the beaver, the ant-hill, and the bird's-nest, are, in this consideration, a part of Nature's works. They grow out of that nature, and the laws by which it operates and by which it is developed. Thus the languages, the architecture, cities, the codes of laws, the systems of religion, the fashions of men, their institutions and civilizations, their progressions in arts, wisdom, and virtue, considered in the largest sense, no less than those works over which man has no control, — the rising and setting sun, the flowing tides, the seasons, the rolling earth, — are but one great Nature; all are but the artistic design and work of God, in the creation of this and of all other worlds. Thus it will be seen that the field chosen by Mr. Graham for these lectures is sufficiently ample, and the subjects indicated may be considered as inexhaustible.

The works of God appear to be the proper study

of mankind. By this they learn the character and attributes of the Creator; they harmonize their lives thereby; they learn skill and wisdom to fulfil their duties over such of God's works as are, either in whole or in part, under their control, and which constitute their art, and make up the sum of *their* work on earth; enabling them, in their degree, to fulfil that sublime precept "be ye perfect, even as your Father which is in heaven is perfect."

It is to the profound thinkers, to men of science, that we must principally look for these lectures. Such we are proud to know abound in our country, who have extended their studies in various directions far into the domain of Nature.

It is with great satisfaction we lay before the public this course of lectures by Professor Agassiz. They have been somewhat delayed, owing, no doubt, to his various scientific engagements and the complexity of his researches.

PETER G. TAYLOR, President.
OLIVER HULL, Vice-President.
THOMAS ROWE, Treasurer.
JOHN W. PRAY, Secretary.
THOMAS WOODWARD,
WM. EVERDELL, Jr.,
ALFRED M. WOOD,
GEORGE KISSAM,
CHARLES H. BAXTER,   } Directors.
DUNCAN LITTLEJOHN,
JESSE C. SMITH,
HENRY W. HERRICK,
AUSTIN MELVIN,

GRAHAM LECTURES.

RELIGION AND NATURAL HISTORY.

LECTURE I.

FOUR DIFFERENT PLANS OF STRUCTURE AMONG ANIMALS.

Ladies and Gentlemen: — You are aware that in conformity with the foundation of this course of Lectures, of which mine are to form a part, a special object is assigned to them; and considering the position in which I have been reported to stand with reference to my opinions in matters of science, I feel bound in frankness to explain to you, before I proceed to my subject, in what light I view the study of nature, and in what way it may promote the object that these lectures are especially intended to further.

The study of nature has one great object which fairly comes within the scope of the foundation of this course of lectures; it is to trace the connection between all created beings, to discover, if possible, the plan according to which they have been created, and to search out their relation to the great Author.

But if science is to contribute its share to the recognition of the existence of God, — if it is to lead the way to Him, from the study of His works, that study must be independent of every other influence; and he who undertakes to state what science has developed with reference to this question must not allow other and antecedent considerations or convictions to interfere. Hence the necessity I feel of presenting to you the results of science in this unbiased spirit.

I know that I have been considered by many persons an infidel, because I have not taken for my guidance in the study of science the dictum of certain creeds. But science cannot submit to dictation, it must build up what it seeks upon the premises which it finds. Let us be content if the results lead to the same conclusion; we shall stand then in the position of one who, having been brought up in the religion of his parents, and having been led astray by doubts, has at length, under the influence of a better frame of mind and of sober thought, come to reconsider the basis of his doubts, and by laborious investigation has returned to the faith he had forsaken.

Such is the position of science. It is the questioning, the doubting element in human progress; and when that has gone far enough, it begins the work of reconstruction in such a way as will never harm true religion, or cause any reasonable apprehension to the real and sincere Christian. Such is my conviction; and while I am considered on one side as an infidel, and decried on the other, in scientific circles, as a bigot, as one who follows the lead of a creed rather than that of science, I feel bound to say that I am neither; and that, if you will receive these lectures in the simplicity with which I offer them, you will find I have not deceived you.

I shall of necessity limit my remarks to the consideration of the animal kingdom, because it is the branch of Natural History with which I am most familiar, and respecting which I can present views which are the result of my own investigation, during a lifetime devoted to the contemplation of nature. The importance of the study of the animal kingdom with reference to its manifestation of the power, wisdom, and goodness of God, is very great. But this is shown only as we advance in the knowledge of the phenomena presented by the animal kingdom.

At this moment natural history can show not only that there is a plan in the creation of the animal kingdom, but that the plan has been preconceived, has been laid out in the course of time, and executed with the definite object of introducing man upon the earth. When naturalists first approached the study of the animal world, they could hardly recognize any system among animals, or ascertain the vaguest relation between them, on account of their extraordinary diversity and their diffusion over the whole surface of the earth, which rendered it extremely difficult to get access to all the different representatives of the animal kingdom. Hence, as a natural consequence, their first study was directed to external appearances. They endeavored to classify animals according to their most obvious resemblances and differences. They put all the aquatic animals in one division under the name of Fish, and arranged all the terrestrial animals under a few classes. Those without legs they called Reptiles; those with wings, Birds; those that walked on all-fours, Quadrupeds; and those who walked on two feet, Bipeds. Next, further distinctions were recognized, as bringing forth living young or laying eggs, breathing through gills or through lungs, and such most obvious features. But such classifications brought together a variety of beings which did not belong together, showing that the method of classification was not natural.

Then followed, in the further study, a survey of the structure of animals. This was a better guide for classification, and anatomy became the foundation of the systems of Zoölogy. Animals were arranged according to their structure, and the moment that key-note was struck an immense progress began. Very soon it appeared that internal structure brought animals much more closely together, according to their real affinity, than features in their external appearance; and following the lead of Cuvier, since the beginning of this century, we have made such progress that we may safely say that at this moment the general affinities or relations among the greater number of animals are satisfactorily known, and that the whole animal kingdom is now classified in a manner not likely to undergo any great modification, except in the details.

But yet one thought lingered in these classifications which was not true to nature. Those who presented them to the world supposed that they exhibited systems devised by themselves, methods of their own ingenuity; and hence they were put forth as the system of this or that naturalist.

But while comparing the different systems with one another, it became apparent to those who sought to trace the source of their resemblances, that, however ingenious, they all agreed in certain respects, and that the discrepancies between the different systems tended more and more to agreement, leading to the conclusion, that, after all, these systems might not be the inventions of their supposed authors, but only their different readings of the systems really existing in nature, and that all the classifications might perhaps be only different translations of one great system. And the moment this suggested itself it became self-evident that the work of naturalists, instead of consisting of ingenious devices for classification, was henceforth to consist only in an attempt to read more and more accurately a work in which they had no part, a work which displayed the thought of a mind more comprehensive than their own, which called into existence the various beings that we see around us, and established their classification. It is now, therefore, the task of the naturalist to read the thoughts of that mind as expressed in the living realities that surround us; and the more we give up our own conceit in this work, the less selfish we become, the more we shall discern, the deeper we shall read, and the nearer shall we come to nature. It is to this attempt at presenting a translation, made conscientiously after reading this plan, that I ask your attention at this time. I shall endeavor to show you that there is really a plan — a thoughtful plan — a plan which may be read — in the relations which you and I, and all living beings scattered over the surface of our earth, hold to one another.

In another lecture I shall attempt to prove that the animal world which now exists on the globe is not the same that existed in earlier times, and I shall endeavor to point out their relations to each other, to show you when these animals were first called into existence, how they have followed one another, and what has been the thought running through the succession of creations, from the beginning, down to the time when, as the crowning act of the Creator, man was placed on the earth at the head of creation.

It was Cuvier who first recognized the fact that the animal kingdom is constructed upon a plan, though that plan did not necessarily imply, according to Cuvier, a conception by an intelligent author. Cuvier only recognized certain structural complications among animals, which brought them together in accordance with the resemblance of their combinations. He recognized four such plans, and showed that all animals, however diversified, are built upon these four plans; and all investigation since that time has only confirmed his discovery. It is true, some naturalists have attempted to show that there are more than four plans, some maintaining that there are five or six, others, even as many as seven or eight; but I think we shall find that they have confounded two distinct ideas, — that of plan with complication of structure; in other words, that, as the same plan is susceptible of several modes of execution, it is therefore possible to confound the mode of execution with the plan; and this is the mistake into which I think these naturalists have fallen.

In the present lecture I propose to lay before you the facts of the case, and not merely to speak of my own views with reference to them; I shall begin therefore by explaining to you these four plans of structure, according to which all animals have been created. They correspond to the four great divisions of animals, known under the name of Radiates, Mollusks, Articulates, and Vertebrates. These four great divisions have been classified into certain species, according to their general resemblance. For more than twenty centuries, ever since the time of Aristotle, all animals having a backbone have been included under the name of Vertebrates. Quadrupeds, Birds, Reptiles, and Fishes all have a solid internal frame known as the vertebral column, or in common parlance, the backbone, surrounded by flesh, within which are cavities containing all their organs, and are, therefore, called Vertebrates. This natural division in the animal kingdom was long known, before it was understood that they were all built upon the same plan; and it requires no small amount of anatomical knowledge to demonstrate this fact. It requires as much practical knowledge, in dealing with facts of anatomy, to trace the relations of one family with another, as it does in mathematics to deal with intricate problems. The anatomist, unless he be very skilful, will fail to furnish the demonstration that the Fish is built on the same plan as the Snake, the Snake on the same plan as the Bird, the Bird on the same plan as the Cat, the Cat on the same plan as Man, and that between all these animals there is no difference as regards the general plan of structure. To furnish that demonstration conclusively is now the aim of anatomical science. The demonstration is possible, and even easy, in its general outline.

Another of the four plans embraces the animals known as Articulates. To this division we refer all the host of Insects, from Beetles, Bugs, Butterflies, Flies, and the great variety of winged animals with six legs, to Crabs, Lobsters, &c., which have a larger number of locomotive appendages, and even down to Worms. All these belong to one and the same division, and are easily recognized by their general appearance; they have rings on the surface of the body, movable one upon the other, and jointed legs, projecting from the sides of these rings. Their relation to one another is thus made very plain by a few general features. But in order to demonstrate that the Crab, the Lobster, the Butterfly, and the Worm are all built on the same plan, it is necessary to show a form of structure capable of transformation into another, and requires great anatomical knowledge.

The third group or division is that known by the name of Mollusks. In it are included those soft-bodied animals, which, like the Oyster, Clam, Snail, and Slug, have a body capable of great expansion and contraction, generally covered with a shelly envelope, or outer coating, of hard substance. Between these animals we find as great a diversity as in the Insect world or among Vertebrates; and yet they are all built upon one and the same plan, and each structure may be transformed into the other without altering in any degree the structural elements.

The last group is called Radiates; the name suggests that they are animals whose parts diverge like rays from a common centre; to this group belong Star-fishes, Sea-urchins, Jelly-fishes, or Sun-fishes, also Corals and animals of that character; and they are built, as we shall see, upon a plan totally distinct and different from that of the other three.

If there be anything which can satisfy us of the working of an all-powerful and comprehensive mind, it is the faculty of expressing one and the same thought in the most diversified forms, and varying the forms of expression in a multitude of ways, so that the thought may be made familiar.

This we have in the animal kingdom, to a degree that challenges our powers of expression, and baffles our most comprehensive modes of thought. Conceive for a moment the whole animal kingdom, consisting of hundreds of thousands of different kinds of beings, constructed only on four different plans. Each one of these plans must therefore necessarily be expressed in thousands of different ways, and it has cost all the thought bestowed by men upon the study of nature till now, to reach the idea that animals are all built upon these four plans.

I proceed, in presenting the characteristics of these plans, to consider the Radiates as the simplest and most easily appreciated.

The Star-fish has its mouth in the centre, from which radiate five arms in different directions. All the organs are arranged like rays in the direction of the arms. The organs of locomotion project from five furrows, on the lower surface of the arms, with suckers at their extremity, with which

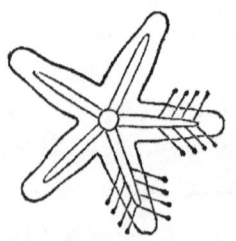

the animal attaches itself to the rocks, or moves about. At the end of each ray, there is an eye. In the centre there is a large circular stomach, projecting into the arms. A nervous ring encircles the mouth, from which rise fine threads, extending to the extremity of the rays and reaching the eyes.

This structure, so far as the plan is concerned,

DIFFERENT PLANS OF STRUCTURE AMONG ANIMALS. 11

is common to all radiate animals; but so far as relates to the mode of execution, it is different in each class. If we examine, for instance, the Sea-anemone, we find that the body is like a sac, from the upper edge of which project feelers in every direction; in the centre we have an opening, leading into an internal cavity, and from that opening hangs another sac internally, which is the digestive cavity. This second sac has an opening in the bottom, which leads into the main cavity of the body, and the main cavity itself is divided by vertical partitions into a number of chambers. On the edge of these chambers, we have bunches of eggs, which in the Star-fish are placed on the sides of the rays.

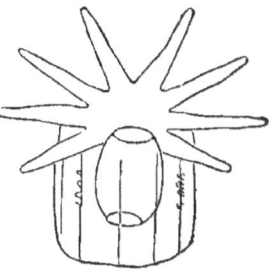

In the Sea-anemone the main cavity, formed by the outer sac, is divided by partitions into chambers, so that you may compare the internal structure to a circular room, divided up into stalls by vertical walls or partitions, leaving, however, free communication in the centre. From each of these chambers projects one of the feelers or arms, which are hollow and communicate with the chamber. In the Sea-anemone, therefore, the idea of radiation, as a plan, is just as obvious as in the Star-fish; but you see at once that the mode of execution is totally different.

Now let us take the Jelly-fish. Seen in profile it is like a hemispherical body of gelatinous substance, from the margin of which hang a number of feelers. 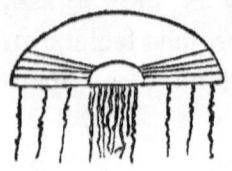 In the centre there is a cavity from which project narrow tubes extending to the edge, and from that central cavity other appendages hang. Looking at the animal from above, we find that the gelatinous substance occupies the whole of this disc. But there is a central cavity hollowed or scooped out of this mass of narrow channels extending to the 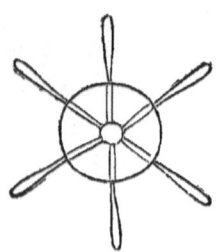 periphery. They may be small or they may be ramified before reaching the periphery, according to the variety of these animals. From the periphery hang these feelers; if there are only four radiating channels there may be but four of these threads; if there are many there will be a corresponding number of feelers. Here, instead of an animal which is, as it were, all hollow, with only a thin wall outside and thin partitions dividing the main cavity into a number of chambers, we have an animal which is all substance. The Jelly-fish has only a small excavation in the centre, with very small tubes extending to the periphery. Yet you see that the general arrangement is the same.

Here, then, is an exhibition of thought in the mode of execution; the differences in structure are only different expressions of the same thought of radiation pervading this group of animals.

You will see at once that each mode of execution leads into the others. Suppose instead of a thin partition there had been a thick one; by increasing the thickness of the partitions, there will be left only a narrow channel between them; and if we fold them above and below, there 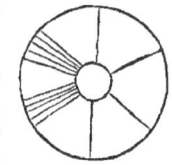 will be only a narrow tube left. So that, after all, the structure is upon the same plan, but in one case broad walls have been made, leaving only thin tubes between them, while in another case thin walls have been made, dividing the cavity into wide chambers. Only a thinking power could devise such a plan; it is not the result of chance. Such close relations under the same circumstances show a power to overcome physical and local influences. For these animals live side by side on the same rock. You cannot visit a single coral reef without finding on its surface thousands of Starfishes, hundreds of Jelly-fishes, almost as soft as the water, Polyps and Corals innumerable, all in the same element and locality, and therefore under the same influences. How can they exist there side by side, except by a higher power than the forces which are active in the sheet of water? Is there not some other cause for their diversity than the influence of heat, light, moisture, and soil combined? One combination certainly cannot produce such diverse results, such different structures upon the same plan. It must be mind acting among these elements, making them subservient to its purpose, and not the elements themselves working out higher combinations of structure.

The next plan of structure is that expressed in the organization of animals like the Oyster, Snail, Cuttle-fish, and the like. Let us take, as an example, the Oyster. If we remove one of the valves, we see the whole surface of the animal protected by a living skin; that skin is attached to the shell. The edge of the shell is provided all around with innumerable little fringes, which are in unceasing action, and are the means by which a constant renewal of water takes place around the animal. In the middle region there is a circular speck of a somewhat tougher substance, which is commonly called the eye of the oyster. It is a bundle of flesh extending between the two valves. ' In a view of the oyster edgewise, we see that the eye, so called, is a bundle of flesh, the fibres of which are attached to the inner surface of the two walls. These fibres by contraction shut the oyster. There is between the hinge, or at the place where the two .valves are united, an elastic ligament. When the two valves are brought together, this elastic ligament is compressed, and the moment the muscular fibres relax, its elasticity forces the two valves apart, and so the oyster gapes. This antagonism is an active means of locomotion. The oyster itself is not capable of changing its place except by means of the fringes around its skin or mantle. The fleshy bundle which extends across the two valves passes through the skin. Below that, there is a fringe

DIFFERENT PLANS OF STRUCTURE AMONG ANIMALS. 15

which extends all around in parallel lines, beyond which, when examined carefully, are found to exist the parallel blood-vessels which form the gills. This fringe is nothing but a respiratory organ. Of these there are two, one parallel with the other, and they exist chiefly in the lower part of the animal, within the mantle, extending to the cross muscle. These fold around the mouth, from which rises a canal, terminating, after sharp convolutions, at the other end of the body; and all this is surrounded by a large mass of liver, which is the green spot you find in the blunt end of the oyster. A ring of nerve extends around the alimentary canal, with a swelling above, and one below, from which sensitive parts go to every organ of the body.

What is particularly striking in this structure, and entirely different from the plan of the Radiates, is, that everything is symmetrical on the two sides of the body along the longitudinal axis, at one end of which is the mouth. In neither the Sea-anemone, the Jelly-fish, nor the Star-fish, do we find anything of the kind. These have a mouth in the centre, from which all parts radiate. There is nothing that can be considered as passing through the animal longitudinally, dividing it into equal parts; but in the oyster we have symmetrical parts arranged on the two sides of the axis, with the mouth at one end. It is particularly characteristic of this Mollusk, that the weight of structure, as it were, is thrown on the sides, so much so, that, if you would examine it to advantage, you should place it on its side. You must lay the oyster flat to see all its characteristic features, and not look at it in profile. You take off one valve and raise one of the hanging curtains or membranes, and thus expose the whole side of the body with all the organs. The load of life is thrown on the sides, though the arrangement of parts is bilateral and symmetrical along the longitudinal axis. That is a peculiarity of this plan of structure.

These animals, moreover, have a soft body, capable of great contraction and expansion, and hence the very appropriate name of Mollusks.

It would be easy to show that between the Oyster and the Snail there is the closest resemblance, and that between the Snail and the Oyster there is no difference in the plan of structure, but only in the proportion and prominence of certain parts. The same may be shown of the Cuttle-fish, or any other animal belonging to the division of Mollusks.

The third group of animals is that of the Articulates, the most prominent representatives of which, as I have already stated, are Insects, Beetles, Crabs, Lobsters, and Worms. Let me take for illustration the Insect: the body is divided into three distinct regions, separated by transverse folds into a number of rings movable one upon the other, forming an articulated cylinder. These rings are so combined as to differ slightly in shape, some being more movable than others, so that the anterior portion, called the head, moves readily on the middle region, and the middle region on the posterior part. And yet all the rings are movable one upon the other. This articulated cylinder has only one cavity, in which are embraced all the organs of life. But these organs are singularly arranged. The breathing organs are always on the side; there is no trace of a breathing organ in what is called the head. Each ring has a breathing-hole on the side, so that if we cut the body longitudinally, we find the lateral holes communicating with a tube, with branches through the whole cavity, filling it with ramifications. The great similarity of Insects is due to the extreme development of their respiratory organs, of which there is one pair to each ring all the way along the body, with ramifications throughout the whole structure.

In addition, they have an alimentary canal, extending through the centre of the cylinder like a tube. Then there is the heart in the upper part of the body, and the nervous system, which consists of one or more swellings in the chest, and one for each ring, along the whole region, so that the nervous system is in the under part of the body between the legs and just above the part where the legs join the side. You see what a singular combination of parts is presented, — the heart on the dorsal part, the nervous system on the ventral, the respiratory organs on the side, and the limbs, when they are only legs, on the lower side of the body, and wings, if there are any, on the upper side. It is a totally different plan of structure from that of the Mollusks or Radiates. It is true these animals are symmetrical; their parts are arranged in pairs on the two sides of the longitudinal axis as in Mollusks; but see the difference. In Mollusks the whole body is one mass, capable of expansion and contraction without a sign of a joint or articulation; while the essential characteristic of Articulates is, that the body is jointed, that every region is movable on the next region, and that there are numerous jointed appendages to some of the rings.

The Crab and Lobster have the same structure, only the wings are wanting, and they have respiratory gills instead of breathing-holes. But the gills are in the same position; they are connected with the sides of these movable lateral cylinders, and therefore the plan of structure is the same.

And so it is with the Worm. It has an articulated cylinder, and if it differs from the Insect, it is only in this, that the rings are all of the same dimensions and proportions, and none of them are connected more closely than others. So that it is only a slight difference in the mode of execution. The locomotive appendages are of a rudimentary character, being reduced to stiff bristles. The respiratory organs are on the side, the organs of circulation along the back, and the nervous system on the lower side. There is no doubt, therefore, that Worms belong to the same type and are built on the same plan as the Crab, the Lobster, the Insect, and all the Crustacea. And to see that this

DIFFERENT PLANS OF STRUCTURE AMONG ANIMALS. 19

is so, we need only look at the growth of some of these animals. What is the caterpillar? It is a young butterfly; it is an animal hatched from the egg of the butterfly. At first it has the worm-like structure; in the next stage of development it casts its skin, shortens and widens its dimensions, opens its joints, which were at first uniform, and forms two regions, like the crab or lobster. Then it remains immovable for some time, takes no food, and passes into what is called the chrysalis state. But it is the same animal in its various stages of progress or growth. In the chrysalis state, it has all the essential features of the lobster, and yet it is a middle-aged butterfly. Then in its final transformation it casts off that skin, the legs and wings are developed, and it becomes a perfect insect. What more positive demonstration could you have that all these animals are built upon one and the same plan?

So we must admit, that, however different in appearance these various animals may be, it is only a difference in the mode of execution, and that they are all formed on one and the same plan, which is different from that of the Mollusks or Radiates.

Let us now pass to the fourth plan, that of the Vertebrates, embracing Fishes, Reptiles, Birds, and Mammals, the latter including Man, at the head of the animal kingdom. The essential peculiarity of these animals is that they have a backbone; but what is more important, they have a double structure, — a double symmetry, — all their parts being represented above and below upon the body. Take, for example, the Fish. The body consists of a backbone, made up of a number of solid pieces. Above each piece rises an arch, which terminates in a spine along the middle line of the back. That arch encloses a cavity. Then below is a similar arch enclosing another cavity. We call these arches the ribs, and we call the spines the backbone. You see that there is a double structure, — one arch above enclosing a cavity, and another below enclosing another cavity, around which are masses of flesh with a bony frame, and over the whole is stretched the skin.

The upper cavity contains only one set of organs, and not all the organs, as we find them contained in the cavity of the Articulates, Mollusks, and Radiates. In the upper cavity are contained only the sensitive organs, which establish the relations between the body and the surrounding world, and through which impressions from the surrounding world reach the animal. That is to say, we have in this cavity the centre of the nervous system, — in the head, the brain; and along the backbone, the spinal marrow. To the brain are attached the organs of sense, and to the spinal marrow the nerves.

In the other cavity, on the contrary, are contained all the organs through which life is maintained, the organs which only go to sustain the animal in its normal condition, namely, digestion, respiration, and circulation, which have nothing to do with the activity proper of the animal. This double symmetry or division of parts is the same in all vertebrate animals, from the Fish up to Man.

As time will not permit me to go into a full explanation of the similarity of structure in this group of animals, I will take, as an extreme illustration, the arm of a man and the fin of a fish. I will endeavor to show you that they are not only built upon the same plan, but the parts are combined together in the same way and almost in the same number. The difference between these animals is essentially this: The head may be more or less isolated from the body by a contraction behind the skull, and the tail may be reduced to a mere tapering point; the extremities may be more projecting, and so jointed as to terminate in the shape of limbs. The covering also may be different; in the fish it may be scales, in the bird feathers, and in the quadruped hair. If you examine the early growth of the scales in the young fish, of the feathers in the young bird, and of the hair in the young quadruped, you will see scarcely any difference. In the penguin, for instance, the feathers in their early growth are so similar to the scales of fishes, that it is difficult to detect the difference. It is only when we view things in their extremes, that we are struck with their differences. We give them certain names, and think that there is no relation between them, when in reality, if we trace them in a series, we find the closest resemblance, as in the case of the arm of Man and the fin of a Fish, which I will now explain to you.

The limb of any quadruped, the arm of man, in particular, consists first of a broad triangular bone which is called the shoulder-blade, from which projects another bone called the collar-bone. Then we have the upper arm-bone, which extends to the elbow. Then there are two parallel bones, extending from the elbow to the wrist. Then eight bones which form the wrist; then five which form the palm. Then we have the thumb with two joints, and the fingers with three joints each. If, without attempting to imitate the human hand, we represent the bones only by dots, the resemblance to the foot of an animal is apparent. Of course the resemblance is not great. It is only when forms are minutely copied in detail, that you obtain the human hand; the moment you fail to produce a detailed outline, and make only a general one, the resemblance to the foot and arm of the quadruped is readily seen. But in the Fish the difference is more striking. There we have the forearm, composed of two bones, one of which is broader than the other. On that are articulated four little parallel bones, and to these we have attached a large fin, consisting of a number of bones united by a web, and each of these bones presenting a number of articulations. But let us examine these ramifications. All animals have not five fingers, some have four, some three, and others two. The horse walks on tiptoe, — on one finger, — owing to the peculiar structure of the foot. But there are other animals in which the number of fingers is much larger. Among Reptiles, instead of three joints of the finger, there is a larger number, and instead of five, there are six fingers. Occasionally in the human family there have been individuals with six fingers. Among animals this is frequent. In Fishes it goes on increasing from six to twenty. So that all these small bones are only so many fingers, each divided into a great number of joints. But after all it is only the spreading hand in which all the joints are united by a web; if we go a step beyond, we find the four bones corresponding to the wrist, then two broad bones which correspond to the forearm; then a very short upper arm-bone; then close to the shoulder a collar-bone and shoulder-blade, exactly as in man. So that the correspondence is complete; it is only another mode of execution, — a modification by the all-skilful Architect of the same plan.

We can therefore say, assuming that mere comparisons like those I have presented to you can be made, that, however diversified the animal kingdom may be, the beings are all constructed upon four plans only. There are none on the earth that we cannot easily refer to one of those four plans.

In my next lecture I shall attempt to show you the classes into which these animals are divided, according to the various modes of execution, and what is the relative standing of the classes to one another, in order that we may be able to examine the question how far the animal kingdom forms one single series from the lowest to the highest.

LECTURE II.

RELATIVE STANDING OR GRADATION OF THE ANIMAL KINGDOM.

Ladies and Gentlemen: — I propose to-night to lay before you the subject of the relative standing or gradation of the animal kingdom. It is one which has long engaged the attention of naturalists. The favorite idea has been that all living beings form an unbroken natural chain from the lowest to the highest, or, as the phrase has been, "from the Monad to Man." This view has been greatly modified in the progress of investigation. It was, as it were, a theoretical view in the beginning, started by general impression, and not at once submitted to the severe scrutiny of a careful investigation; and when that was made, this idea was found not to be true in the form in which it had been stated. And yet the investigation has led to some interesting results; as, for instance, the discovery of the limits within which there is a gradation or relative rank, and beyond which there is not.

This view of an unbroken continuity of the animal kingdom from the Monad to Man, was started before it became known that all animals are constructed upon four different plans, so that the differences of structure with reference to these plans was no obstacle to the formation of a supposed series or chain. The discovery of the four plans has in a great measure checked that theory, though at this moment it is reviving in connection with questions which I shall hereafter consider, especially with reference to the question now so extensively discussed among scientific men and philosophers, of the origin of species, of which I shall have something to say presently.

But before considering that question, we should become familiar with the broad, comprehensive principles lying at the foundation of all natural science. I have already shown you that there are four great plans upon which all animals are built, and that those plans are essentially different, so much so that it is difficult to conceive how a transition from one to the other could be made. But before this knowledge was reached so that it could be demonstrated to ordinary minds, there was a general and vague impression of a gradation in the animal kingdom. It was evident that man was superior to the rest of the animal creation. His structure presented greater complication, and there was a remarkable superiority in some of the most essential organs, especially in the development of the brain. It was already shown by Kæmpfer, that among all the animals the Monkey, especially the Orang-outang, came nearest to man, and it appeared as if from man there was a gradual descent to the lowest animal creation. From air-breathing, aquatic animals there appeared to be a transition to whales, which in those days were erroneously supposed to be fishes, and thence to gill-breathing fishes. With reference to the lower orders of creation, it seemed hardly worth while to consider them, as the series was so plain among the higher beings; and as it was supposed that sooner or later the same gradation would be discovered among the inferior animals, the principle was taken for granted. The bulk of those lower animals were according to this view reduced to a few types. Worms were thought to form a sort of connecting link with snails, eels, and the elongated Vertebrates. With the Worms were associated shell-fishes and the like, and the class was extended so as to include some animals which were supposed to be more simple. The microscope not having applied its severe test of scrutiny to the structure of the so-called animalcules, they were considered as mere animal globules, capable of moving, of feeding, and sustaining existence. Hence it seemed a legitimate conclusion that there was a single uniform series, beginning with the lowest beings, called Monads, and rising to the highest, which was Man.

But when, by a deeper and deeper study of the structure of a larger and larger number of animals, it was attempted to complete the links in the series, or to test more accurately their relative rank, difficulties began to arise, and doubts to be entertained as to the real existence of such a series, until at length the conviction became prevalent, that, instead of such a single series, there were many series; and the question now is, within what limits do these series obtain, and where do we fail to find them in steady progression?

I will in this discourse consider the facts bearing upon this question. An insight into this subject is absolutely necessary, before we proceed to consider the order in which living beings have been introduced upon the earth. For, of the two theories which are generally entertained now, namely, either that all animal species have been developed from the lowest to the highest, one from the other, or that they have been created as they exist, one or the other must be false. And if we have no single series of progression from the lower to the higher, then the first of these views becomes less probable than it otherwise would appear.

But the consideration of this question involves the necessity of a further examination of the classification of animals. And I will begin by showing you in what way the Vertebrates — that type being more universally known, and being also the highest of the four primary divisions — are subdivided into classes.

But I may, perhaps, be able to make more clear the nature of this investigation, if, instead of proceeding at once to the subject, I introduce a comparison from the study of languages. This I may be pardoned for doing at this time, inasmuch as there is hardly any one who does not know some language besides his mother-tongue. Now, if I can show you first what philologists have been aiming at, I shall be able to show you more readily what naturalists have been aiming at.

Take the common word "father." In the Latin and Greek it is "pater," in the French "père," in the Italian "padre," and in the German "vater." At first it may perhaps seem that there is not a very close connection of sound between these words. But it is found that there is a law prevailing, which the philologists in Germany, and especially the Messrs. Grimm, have detected, in the gradual transformation of letters, which is as unerring in the progress of human development as any other law of nature. It is found that the hard letters are the older, and that they have gradually softened in the course of time. Thus *b*, *p*, and *ph*, which are equal to *f* or *v* in pronunciation, have succeeded each other very much in the same manner as *d*, *t*, and *th*, or as *g*, *k*, and *ch*; and all the words in which these letters occur among the ancient, and especially the Southern languages, have at later periods been transformed into other languages. Thus the Latin and Greek "pater" becomes in the Gothic "phthar," in the German, "vater," in the Italian "padre," in the French (leaving off the "pa") "père," and in the English "father." All these words are therefore so intimately allied that it is impossible to mistake their genetic connection; and if you have occasionally heard people speak two languages, you will see how natural it is for them to pass from one to the other. The German does not pronounce the "th" as we do, who have learned to pronounce it a little better than Europeans. There are others who pronounce *p* like *b*. All that is the natural result of local association.

If I were to dwell upon this subject, I might go a step farther, and show you that the Hebrew word "baré" (to create, to make, to prepare) is the same as the word "bear" (to bring forth); and that here again we have the root of the word "pater" (the originator, or father); and still further, that a host of derivatives, such as "parent," &c., are brought into this association, merely undergoing a slight change of outer form or intonation, showing the genetic connection of all these words.

Now transfer this idea to the study of the structure of animals. Conceive a primary idea at the foundation of all the animal structures, according to which they have been called into existence. If we can obtain an insight into their various structures, we shall be able to detect the affinities among animals, just as the etymologist discovers the internal relations among words. It is a kind of etymological study of the structure of the animal kingdom. It is going back beyond the mere external appearance, and seeking for the origin of things. And we may perhaps be able to see how far that goes, when, besides the analysis of these structures, we proceed to consider the question of the first introduction of animals upon the surface of the earth in early geological epochs, to which I shall advert in my next and following lectures.

Among Vertebrates we have, as they are generally described, four classes, namely, Fishes, Reptiles, Birds, and Mammals. On the average they stand one above the other in the order in which I have named them, Fishes being the lowest and Mammals the highest. This rank is readily conceded to the latter. It is also readily perceived that Birds must stand above Reptiles, a class of animals in which there is a simpler circulation; and there is no reason why we should not place Fishes the lowest, when we remember that they are aquatic animals, breathing through gills, destitute of the power of coming on land, and in every respect of structure inferior to the other three classes. For a century past, no naturalist has doubted the relative rank of the four classes in the order in which I have given them.

But when the question is asked, whether, beginning with the lowest species of Fishes and ascending to the highest, we can thence make a transition to the lowest Reptile, in an ascending series, and then, rising to the highest Reptile, pass to the lowest Bird, thence to the highest, and then again pass from the highest Bird to the lowest Mammal, and so on up to Man, we are placed in the same embarrassing position in which naturalists were when they attempted to arrange all animals in one single series, before the discovery of the four distinct divisions. We cannot make the passage from one group to the other, without doing violence to the internal arrangement of those several groups. For though it may be acknowledged that Fishes as a class are the lowest, Reptiles next, Birds next, and

Mammals highest, it does not follow that we can string them together so as to place Reptiles entirely as a class above Fishes, Birds entirely above Reptiles, and Mammals entirely above Birds. When we come to consider Fishes among themselves, it appears difficult to determine which species is lowest and which is highest. Anatomy seems to have settled the fact pretty nearly that Sharks and Skates are superior to the other species. Their limbs are more developed, and their anterior region has acquired a preponderance not noticed among ordinary bony and scaly fishes. If we acknowledge this position to be correct, then we must place the common fishes, and among them the Lamper-eel, which is very simple in its structure, at the opposite end. Then we have a series beginning with the Lamper-eel, next would come the ordinary bony fishes, then those singular fishes of our Western waters, the Gar-pikes, which bear a strange resemblance to Reptiles, and lastly Sharks and Skates at the head of the series. This would seem to be the best arrangement that can be proposed, according to our present knowledge of these animals. But a closer scrutiny will show that such an arrangement cannot be maintained, though it is as nearly in accordance with their rank as we can make it.

We next pass to Reptiles. This group embraces Tortoises, Crocodiles, Lizards, Salamanders, Serpents, Frogs, Toads, and the like. If we examine into their structure, it is at once apparent that there are two classes of Reptiles, which differ widely from one another. One is essentially aquatic; the animals are hatched from eggs, and in the form of tadpoles live like fish in the water, and undergo a succession of very striking changes. They are not born in the form of the adult animal; every Frog or Toad is at first a Tadpole, which has gills like a Fish instead of lungs. Gradually, however, it loses its tail, puts forth legs, develops lungs instead of gills, and assumes the characteristics of an air-breathing animal. The other class, embracing Snakes, Lizards, (not Salamanders, which are erroneously called Lizards, but which belong to the Frog tribe,) the Crocodile, the Alligator, the Tortoise, all have a very different mode of development. Their eggs are like Birds' eggs, and their young when hatched have the form of the adult, with no further change except in growth. They breathe through lungs at birth, like the adult Reptile. You see at once, that there is reason for placing these fish-like Reptiles lowest in the scale, in consequence of their resemblance to Fishes, and the air-breathing Reptiles highest, because in that respect they resemble Birds and Mammals.

This is then the order of Reptiles: Salamanders (commonly called, in the Northern States, Waterdogs) lowest, then terrestrial Salamanders, then Frogs and Toads; and from these we pass at once to the scaly Reptiles. If we attempt to establish an order among the scaly Reptiles, we find that Serpents, on account of the absence of locomotive organs, and on many other accounts, (I have not time to go into the anatomical details,) rank the lowest; then we have Lizards with rudimentary legs, then the true Lizards with limbs very fully developed, then Alligators and Crocodiles, and lastly Tortoises, which are the highest.

And now what series would we have? Salamanders lowest among the aquatic Reptiles, and Frogs the highest. Then from Frogs we would pass to Serpents, which are the lowest among the scaly Reptiles, and from them to Tortoises. So then, if we attempt to arrange our series in connection with Fishes, we have first, Lamper-eels, the lowest of the Fishes, then bony Fishes, next Gar-pikes, and then Sharks or Skates, the highest. From these, attempting to pass to the lowest Reptiles, we should come to Salamanders, then to Frogs, then to Serpents, and lastly to Tortoises. And now we come to Birds. But whatever kind of Bird you select to connect with the Reptiles, you see at once what an awkward transition you make from Turtles to Birds. And yet unquestionably Tortoises are not only the highest among Reptiles, but they come nearest, of the Reptile class, to Birds. And this is a fair example of the violation of natural relations which we commit when we attempt to arrange the whole animal kingdom in a single series.

But let us proceed with this attempted arrangement higher up. Even if we entertain this idea of a passage between Tortoises and Birds, we must take the water Birds as the lowest. If we take those that are hardly able to fly, such as the Penguins, and bring them by the side of the Tortoises, what species of the latter do they resemble most? The sea-Turtles. But it is demonstrated that the land-Turtles are higher than the sea-Turtles. And yet if we attempt the comparison between aquatic Birds and Tortoises, we must make it not with the highest Tortoises, which are land-Turtles, but with the inferior, or sea-Turtles. But having started with the water Birds, which are the lowest, and progressing through the wading, the running, the gallinaceous, the climbing, and the singing Birds, we reach the Birds of prey, which are generally considered the highest. Suppose this to be the true order, or any other; — some have supposed that Parrots stood at the head of Birds, others have placed the singing Birds at the head, others, Ostriches, on account of their size; but I will not discuss these questions; it is enough that through some one species we should attempt to make the transition to the Mammals; — whichever species we select, the transition will be awkward, quite as much so as from the highest Reptile to the lowest Bird. There is no possibility of passing from one class to the other.

And when we attempt to arrange the Mammals among themselves, the awkwardness of such an attempt to establish a single series is still more apparent. The lowest among them are the Whales. Whales are not Fishes; they have lungs, a double circulation, and warm blood; they bring forth living young, and nurse them with milk, like Quadrupeds. They belong to the same class of animals to which we belong, only they are the lowest.

Their structure is more nearly like ours than it is like that of Fishes, to which they bear such a striking external resemblance. These, then, being the lowest of the class of Mammals, if we attempt to connect them with Birds, see what follows. From Whales, Porpoises, or Grampuses, or Blackfish, all of which belong to the same order, we must pass downward either to Birds of prey, singing Birds, Parrots, or Ostriches. Is such a transition possible?

But there is, nevertheless, something true in the idea of gradation, only it is not so simple a matter as those who first propounded it conceived it to be. Instead of a single series, a uniform gradation, we have complicated relations, all embraced, however, by a comprehensive idea. The truth is, that while the Mammals are unquestionably the highest group, they are not in every respect superior to Birds, and while Birds are on the whole superior to Reptiles, there are some Birds that are inferior to some Reptiles. And the same is true of Reptiles and Fishes. On the whole, Fishes are inferior to Reptiles, Reptiles to Birds, and Birds to Mammals. But they do not stand in an unbroken series, one above the other; they may be represented thus:

| Mammals. |
| Birds. |
| Reptiles. |
| Fishes. |

Within each of these classes, there are several parallel series, more or less closely linked together, and which, when compared, present strange correspondences. I will take one example for illustration. I have described purposely in some detail two classes of Reptiles, for the sake of introducing this point. There are fish-like Reptiles that during the early stage of life have gills; they lay a large number of eggs; and undergo successive transformations. They are considered by some as forming a class under the name of Amphibians. The lowest of this type is called the Conger-eel, and is found in the Southern States. It has very rudimentary limbs and large gills, and is known by the name of Syren among naturalists. Then we have the Salamander, in which the limbs are more fully developed, and gills exist only in the early stages. Then we have those animals, such as Frogs and Toads, which, in passing from the Tadpole state lose their tail, acquire limbs, and develop lungs instead of gills. Now among the scaly Reptiles, called Reptiles proper, we have a similar gradation. First, we have the Serpents, the lowest, next the Lizards, and lastly the Tortoises. If you compare these animals, you will find that the Syrens resemble the Serpents, the Salamanders resemble the Lizards, and the Frogs (especially the large tropical animals provided with a shell over their heads) are allied to Tortoises. So we have here two classes, which are strictly parallel in their gradations, having the same features with certain modifications of structure. And such a parallel series you find everywhere among these animals. So that the truth as regards series is this: that instead of one grand uniform succession from lowest to highest, there are certain broken series within the minor groups, in which it is possible to trace more closely and intimately links of subordinate series.

Something of the same kind obtains among the Articulates. The best arrangement of this group reduces the number of classes to three, namely, Worms, Crustacea, and Insects, the latter being the highest. The relative standing of each of these classes is determined by the nature of the complication of their rings. In Insects they are combined in three regions, in Crustacea (Crabs and Lobsters) in two, and in Worms in one. In Insects the posterior region is generally made up of nine joints, the middle region of three, and the anterior region of one; there are limbs attached to the middle region below, wings above, and appendages attached to the head. These animals are superior to Crabs and Worms in consequence of this peculiar arrangement of the rings.

In the Crustacea you have only two regions. The posterior part consists of movable rings, and the anterior part of rings that are not movable. But the anterior part contains long locomotive appendages, called legs, then a number of other appendages in the shape of mouth-pieces, and then long feelers. But the essential characteristic of Crustacea is the division into two regions.

In Worms the essential characteristic is one uniform cylinder. All the rings are of nearly the same size. That fact alone is sufficient to indicate their rank as the lowest of the Articulates, and all further study of their internal structure justifies and confirms that position.

But yet we have a certain number of Insects which are Worm-like in form. Centipedes, for instance, have a body divided by a number of nearly equal joints. But if you ask, why then not refer them to the class of Worms? I answer, for the very obvious reason that their essential organs are those of Insects. They are a lower form of a higher class, but from their external appearance they seem to belong to a lower class.

Again, we have Spiders, which are divided into two regions; the head is not separated from the chest, as it is in Insects generally. But what shows that you cannot refer the Spider to the class of Crabs, nor the Centipede to the class of Worms, is, that all Insects are air-breathing animals, having air-tubes opening upon either side, one pair to each ring, ramifying in the body; and their other anatomical features bind them in the same class together, though the outward form of the Centipede is more uniform than in Insects generally, more like the Worm, and the Spider is externally like the Crab. The Centipede stands in the same relation to Insects as the Whale does to Mammals. The head and chest are not divided by a contraction between them, the limbs are not protruding and free, but are packed together. So we have first, Centipedes, which are lowest among Insects, next Spiders, and then Insects proper, which are the highest.

Hence though we are justified in placing Worms

lowest, we have some Insects which are Worm-like; and though we are justified in placing Crustacea next, we have Spiders which resemble Crustacea externally, but which must be ranked among Insects. Thus the classes are interwoven as it were. When you take the essential features of structure you bring them together; when you consider the external form only, you have yet recognized relations which escape in the other combination. There is, therefore, no more possibility of arranging one continuous series of these animals than of the Vertebrates. For from the highest Worms we should pass to the Crustacea, from the highest Crustacea (the Lobster) to the lowest Insect (the Centipede), from the Centipede to the Spider, and from the Spider to the winged Insect. It is only in the minor divisions that we find a natural gradation — not in the arrangement of the whole group. The classes are superior or inferior to one another on the average, but not by such an arrangement as would place them in one serial order.

If now we proceed to the group of Mollusks, we have the same facts. Mollusks are divided into three classes, namely, Acephala (embracing Clams, Oysters, and the like), Gasteropoda (embracing Snails, Slugs, Periwinkles, and the like), Cephalopoda (embracing Cuttle-fishes, Squids, Nautilæ, and the like); and their relative standing is, Squids and Cuttle-fishes highest, Snails and Slugs next, Clams and Oysters lowest. One single fact of their structure will at once show you their relative standing.

In the Clam we have a proboscis made out of two tubes through which water is introduced into the shell. We have gills on the sides, and other interior organs, and protruding appendages

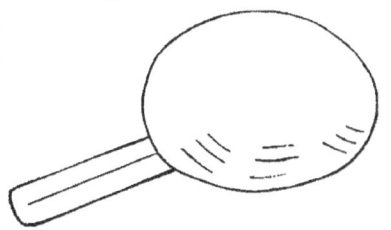

hardly capable of moving. But when we come to the Snail we have the anterior parts more highly developed, and marked by a striking

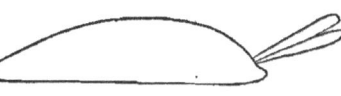

prominence of these appendages. Passing to the Slugs we find the head marked out by a contraction of the body, large eyes, and a number of large appendages arranged around the head, commonly called arms, eight in number. The head is made prominent, while in the Snail it is only slightly indicated,

and in the Clam it is entirely wanting. The name Acephala (headless animal) has really a reference to the total absence of specialization in the anterior region of the body, which should be the head. You see, therefore, that the marking out or defining of that anterior portion indicates the position we should assign to these classes.

Again, if we attempt an arrangement of all these classes in a single series, we find that there are some Acephala which are superior in endowment to certain Gasteropoda; and we find certain Snails and Slugs which are superior to certain

Cephalopoda. So that here we find the same difficulty again in attempting to form a single series, while the minor serial arrangement is apparent.

As regards Radiates, the relative position of the classes is most curious. Radiates are divided into three classes, namely, Echinoderms (the highest, embracing Star-fishes and Sea-urchins), Acalephs (the next, embracing Jelly-fishes and the like), Polyps (the lowest, embracing Corals and the like). This is their natural order, as I explained in the first lecture. But any attempt at a general serial arrangement among them leads to the same difficulty as heretofore. There are in all these classes some animals that are inferior to certain animals of the class below; so that it is only the average bulk or weight of the class which assigns to it its relative position.

A few facts respecting their mode of development will show you how strikingly their features of growth agree with the fact of their relative standing. You know what Jelly-fishes are. Animals of a very soft nature, with a hemispherical form, from the margin of which hang innumerable fringes, and from the centre of which hang long appendages which surround an opening leading to the internal cavity; thence tubes radiate to the periphery. There is nothing like a Polyp here, —no body like a sac with fingers or tentacles around the upper margin, and a central sac leading into a digestive sac below. The structure is quite different; and yet the egg of the Jelly-fish,

GRADATION OF THE ANIMAL KINGDOM. 43

when hatched, (and I have raised too many of them not to know exactly how they are developed,) produces a young animal which is at first a free swimming creature in this form, covered all over with fringes, by which it is set in motion. It has no cavity, but is a homogeneous mass of cells, as closely packed in the interior as at the surface. Now the first change indicated in this little animal is a depression of form at one end, and as that depression is enlarged, the cells, or substance of the animal within, are gradually loosened; the interior becomes less and less solid, and at length entirely liquefied. As soon as that is completed, a communication between this depression and the interior is effected, and the animal has now a main digestive cavity and a mouth. But it still moves freely about. Could you imagine it was anything but a free animal? But that is a mistake. The next step is, the animal attaches itself to the surface of the rock, the point by which it attaches itself flattens and spreads, the upper part of the body widens, and from the corners of that opening begin to protrude four short tubercules, which in a short time lengthen into four long tentacles; then between these four more, opening and leading into the main cavity; so that the young Jelly-fish is now a Polyp-like creature, so much so that it has always

been described as a species of Polyp, until by direct observation, by raising it from the eggs, it was found to be the progeny of the Jelly-fish.

There are two ways in which the Jelly-fishes are developed. First, in the Polyp-like form, until they appear like a long stem with a head at the end and tentacles on the summit, or scattered over the head. Then in course of time certain buds grow out at the side and become gradually more and more prominent, presenting at the same time marked changes. Here are buds which after a certain time drop off, and then the animal has this form, with only a few tentacles. But it is a genuine Jelly-fish developed from the body of the Polyp-like animal, born from the egg. In the other case this little Polyp-like body grows, becomes a long stem surrounded by tentacles, and has a mouth. Presently several contractions arise, from six to fifteen or twenty in number, which grow deeper and deeper until the whole stem is transformed into what looks like a pile of saucers. Through the whole stem there is a central cavity. After these contractions have become very deep, and the discs in the pile have become free along the edge, the hole is widened to a considerable extent, and the top of the pile drops off and dies. Each disc then separates and assumes this outline. From the central cavity eight tubes arise, and extend to the margin of eight 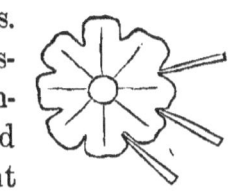 leaves. Presently fringes project from these, and then you have a little Jelly-fish; the resemblance is apparent in every respect, and the animal only requires to grow in order to become a perfect Jelly-fish. Each one of these discs drops off to the number of ten, fifteen, twenty, or more, and becomes a Jelly-fish; so that from one egg is born a Polyp-like animal, which divides up into many animals of an apparently different species. Here the idea of the Jelly-fish is so combined with the idea of the Polyp, that the structure of the latter is made the pattern of growth of the former. If this does not show that the same thought existed in the foundation of the law of this growth, then it does not demonstrate anything. And we have the additional extraordinary feature, that out of one egg are born a large number of individuals. These facts seem almost like fables, and yet they may be tested every day. Yesterday I witnessed the birth of one of these Jelly-fishes. And if you will at the proper season gather the eggs of these animals, often stranded on the beach, and take care of them during the winter, (and they are easily kept,) you will have an opportunity of obtaining an insight into this complicated relation among Jelly-fishes.

Knowing, then, the general classification of the animal kingdom, built up as it is upon four plans, in the order I have described, you will at once see that to attempt to place these four great divisions one above the other, we should have to make the transition in this way. Beginning with Star-fishes and Sea-urchins, which are the highest Radiates, we pass to the lowest Mollusk, represented by the Oyster; then rising to the Squids and Cuttle-fishes, at the head of Mollusks, we make a transition to the lowest of the Articulates, represented by the Worms; then rising to the winged Insects, the Butterfly, for example, we pass to the lowest of the Fishes. So that the moment we have established, by a better insight into the minor affinities of animals, their true order and relative position, the impossibility of establishing a single series becomes evident.

With these facts before you, I shall be able to proceed in my next lecture to discuss more satisfactorily the question, when and in what order of succession have animals been introduced upon the earth?

LECTURE III.

REMOTE ANTIQUITY OF ANIMAL LIFE AS SHOWN IN THE
CORAL REEFS.

Ladies and Gentlemen: — I propose this evening to consider a question, which has long attracted the attention of the world, and led to some angry discussions, but which we must nevertheless face, like all other questions of the day, and meet openly and frankly. The question is simply this: When have the animals inhabiting the surface of the globe been called into existence? There is another question intimately connected with this, which I will take up in my next lecture, namely: In what order have those animals been called into existence?

You see at once that this question involves also the question of the age of the world. For a long time it has been believed that we knew exactly how old the world is, and that there could be no question as to its comparatively recent origin. But this impression has resulted from the fact that the chronology of the world has been confounded with that of the human race. The traditions of the relative existence of man on earth, and the history of the development of nations, have been taken as the chronology by which to measure the age of the universe. Students of nature, however, have found that the standard of human history does not apply to the history of the physical world. There are facts showing that our earth is much older than the existence of man on its surface. I propose, therefore, to consider those facts, as based upon observations of nature.

And while at the outset I set aside all tradition with reference to this question, let me not be understood as supposing that there is any conflict between the narrative of Genesis and the results of scientific investigation. "In the beginning God created the heaven and the earth." When that beginning was, Genesis does not say. That beginning is so far remote from the period when man was called into existence, that the question of the age of man, and of the animals and plants now existing on earth, we may fairly say, has nothing whatever to do with the question of the age of the world.

But to approach the subject under consideration, it is interesting to gather the evidence at hand respecting the time which has elapsed since the animals now existing were created, and also respecting the changes which they may have undergone, in order to obtain light upon the question of the successive introduction of all the varieties of animals now known to exist on earth.

Our domestic animals have always followed man in the progress of civilization. Wherever the traces of civilization are found, there are found also traces of the presence of animals, not only domesticated, but also wild. No civilization has left us more interesting traces in this respect than that of Egypt; on the Egyptian monuments are represented in sculptures and drawings, and in the catacombs are preserved in the shape of mummies, animals, which lived many thousand years ago. Some of those relics, which have come down to us, are unquestionably nearly five thousand years old. They form a very interesting basis, by which to ascertain to what extent animals may change under the different circumstances in which they live. The most careful comparison which has been made between the skeletons of the animals preserved in mummies, and those recently killed in the valley of the Nile, has not shown the slightest difference between them. We have here, therefore, direct and positive evidence that a period of five thousand years does not change the appearance, character, or structure of any living being.

But there is something more to be considered in regard to domestic animals. They are constantly modified under the fostering care of man, and in a way that is peculiar, — very different from what we observe among the wild animals. This feature should be kept in mind whenever the question of the diversity which we observe in nature is under consideration, — especially since there are those who believe that the varieties we find among our domesticated animals, may afford the clue by which to explain the diversity which exists among wild animals. It is an undoubted fact that the differences among domestic animals, which we designate by the term "breeds," are of comparatively recent date. The time when many of them were first introduced is known, the variations are the work of man,— the result of human care, of artificial means. But these differences are not of the same kind as the differences we observe among wild animals.

Let us compare some groups of wild animals with the domestic animals to which they evidently belong. There is roving in the Western prairies a wild kind of bull called the Buffalo. It is a distinct species of that family, not yet domesticated. There are different kinds of wild bulls in Africa; on the northernmost part of this continent there is still another kind called the Musk-ox; there is also another kind on the Table-Lands of Asia. All these wild bulls differ from one another in certain ways: either by the character of the hair, the form of the horns, their proportions, size, or color. But all those of one kind are so similar among themselves, that he who has seen one of a herd of Buffalo has seen them all. And the same is the case with the Musk-ox, the Aurochs of Europe, and the Yak of Asia. The animals of each species have a great uniformity among themselves, though they differ widely from those of the other species.

Now examine different breeds. One of the first results of domestication is to introduce an extraordinary diversity among the different individual heads. There are no two heads alike in a herd of domesticated animals. They differ, not only in color and size, but also in the form or twist of the horns. And these differences are only constant as the animals are kept under the same influences. Breeders of cattle know, that, unless they take care of their herds, all the improvements which domestication has introduced will disappear.

Here, then, we have the evidence that these differences are the work of man, the result of artificial means applied for the purpose of rendering the animals subservient to him; while on the other hand, the differences existing among wild animals are the result of a creative power over which the mind of man has no control. Domesticated animals show us only the amplitude of the pliability of structure in each animal, and in no way the method by which the diversity existing among wild animals can be supposed to have been introduced. Domestication never produces forms which are self-perpetuating, and is therefore in no way an index of the process by which species are produced.

Thus we have two important facts, namely: 1. That a long period of time does not cause differences among animals that are left in their natural condition of existence. 2. That the mind of man alone is capable of producing the extreme differences we observe among the breeds of animals.

Let us now consider the data which are afforded us respecting the age of the animals which now inhabit this globe. There is no field of observation so rich for this purpose as the coral reefs, and among them perhaps none are so instructive as those which skirt the southern extremity of our own country. I will therefore lay before you to-night what I know of the history of the coral reefs of Florida, of their origin and growth.

And in order that I may be well understood, it will be necessary to recapitulate some statements respecting the nature of the animals which build those reefs. Coral reefs are banks or walls of hard material, growing from a definite depth to near the surface of the water. They are entirely the work of marine animals, commonly known by the name of Corals, and belonging to the class which naturalists designate under the name of Polyps. The prevalent idea that corals are the work of a kind of insect, has arisen from a mistaken impression derived from the sight of corals, many of which look very much like a honey-comb, suggesting therefore the idea that an animal somewhat resembling the Bee has produced them. It is not so. Corals are a part of the body of an animal, as much so as our bones are a part of our frame. The corals we see are the solid portion of that animal when alive. In earlier times, when the classification of the animal kingdom was not based upon difference of plan of structure, all animals belonging to the lower classes were referred to two groups, Insects and Worms. In those days these animals were called Worms by some, Insects by others. They are neither. They are built upon the plan of radiation; they consist of a number of equal parts, diverging from a vertical axis, and arranged in a perfectly symmetrical way. They have a central mouth, and a number of feelers surrounding the upper part of the body, which receive the food. They are identical with our Sea-anemones, differing only in this, that they are the hard parts of that animal. I will explain to you presently how these animals, having such hard parts, are at the same time readily movable, and can conceal themselves altogether within those hard parts.

The Sea-anemone is an animal of this character. In the centre it has a mouth with feelers all around. 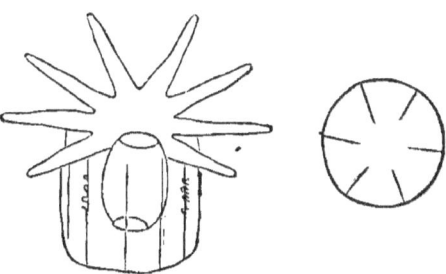 This mouth opens into a sac, which is the digestive cavity. At the bottom of this cavity is a hole, through which the digested food is carried into the main cavity of the body. This main cavity is divided by radiating partitions into a number of chambers communicating with one another at the centre. The partitions are not united at the centre, and therefore the different chambers communicate with the digestive cavity. Such an animal, when soft, is called a Sea-anemone. But let the walls be loaded with limestone and become stiff, then we have a coral. Thus, not only the side-walls become stiff and hard, but each partition also, so that the radiating divisions of the cavity remain visible when the soft parts are decomposed. If you examine a coral attentively, you will perceive that each piece on the surface presents the appearance of a little star, owing to the radiating partitions projecting from the outer hard walls towards the interior. But now the upper part of the animal remains soft and movable, capable of expansion and contraction to such an extent, that, in its fully expanded state, the upper portion may be two, three, or four times higher than that portion which has solid walls; and yet, if the animal is disturbed by the approach of anything, and is apprehensive of danger, it draws in its feelers, contracts itself, and coils up so that the whole of it may be hidden in the slight depression in the centre between these radiating partitions and the outer wall, and there remains nothing visible at the surface but a cup-shaped, hard body. The whole of the soft parts of the animal in that way disappear from sight, and are only discoverable upon careful inspection.

Another fact in regard to these animals. Instead of being single, like the Sea-anemone, Corals are generally compound; that is, they multiply without dividing. The successive generations do not become separated from the parent stock, but go on growing attached thereto. We are so familiar with this phenomenon in plants that we hardly notice it. But when we refer to it as a fact occurring among animals, it strikes us with wonder. The little plant, germinating after throwing out a few leaves for the first year, comes to a stand. It is a little stem with one bud. What is it but an individual grown to that size? The little bud at the head is the next individual, which grows upon the plant the next year. And so on, bud after bud, each of which is an individual growing upon the parent stock, until finally we may have a large tree producing thousands and thousands of buds, no one of which will separate and form a new tree until we cut it off and ingraft it on another plant, which gives us evidence that it is an independent individual. But yet the whole tree is not a single individual, but a community of individuals, growing in close connection and never separating spontaneously. So corals are communities, masses of individuals, growing up in the same way, budding side by side, or dividing in another way, and while dividing or budding or multiplying, remain united together so as to form a larger and larger mass. Around the small body of a Polyp will grow first one bud here, another there, then three or four on this and on the other side, until they form a nest around the parent. Then others will form in the 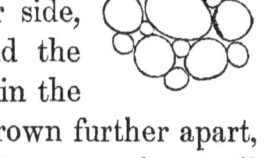 intervals when the first have grown further apart, and will in their turn acquire the same size, until we have a multitude of individuals all united in one mass.

The buds are so arranged that the end has the

form of a hemispherical body. You must have seen some of these masses of coral as they are frequently exhibited in the windows of drug stores, showing a rounded form. But there are others which are like branches divided and subdivided, and presenting the most beautiful ramifications. All this is owing to the manner in which the new individuals unite with one another. We find the same among trees. The buds of the oak are different from the buds of the maple, the willow, or the poplar; so much so that we are enabled to detect, from the mode of ramification, the different kinds of trees even when they are destitute of leaves. So each species of Polyp has its own peculiar mode of budding, branching, and ramifying, giving it a different external appearance; and what we know as leaf corals, brain corals, finger corals, and all the intermediate forms, are only the result of different modes of multiplication or budding. The number of these different species is very great; there are as many different kinds of corals in the seas as there are of fishes or shells, only our attention not being turned so much to the corals, the common vernacular has not supplied us with their different appellations. Naturalists, however, have distinguished many different kinds. They all have peculiar habits and features, and require different positions in the sea. There are those which are only found in shallow waters; others never grow to a height above two fathoms; others are never found in waters which are less than five or six fathoms deep; others grow only in waters at least ten fathoms deep. And these peculiarities are as constant as the differences we observe in the distribution of plants and trees on mountain-slopes, or in the distribution of animals. There is only this difference: that while plants and animals have a range more or less extended, the limits within which corals will grow are very narrow, and the fact of the water being more or less clear is enough either to foster their growth or cause their destruction.

You will see what striking conditions come to bear upon them when you consider that at the level of the water there is one atmosphere pressing upon them; at the depth of thirty-two feet the weight of another atmosphere is added; and at the depth of sixty-four feet there is a pressure equal to three atmospheres. An animal made of such soft and tender materials must be very nicely and evenly adjusted in its structure to be able to bear the pressure of a particular depth and no other. Each kind is as marked in its level as the range of trees, and more so. Whoever has lived at the foot of a mountain-range and has seen vegetation progressing in the spring, may have noticed that certain trees will form horizontal lines along the base of the mountain when spring sets in, resulting from the earlier budding of some plants, and as the season advances these lines of vegetation will rise against the sides of the mountain-slopes. Now if this difference of range exists among trees, how much more definite must be the different degrees of pressure upon animals in the ocean, inhabiting different depths.

We have then two conditions relative to these animals which will bear upon the question of the formation of coral reefs, namely, 1. That these animals are influenced to a great degree by the conditions in which they grow, and are extremely limited in the range they occupy. 2. That they are very different in their structure, one from the other, so much so that one species cannot be mistaken for another.

A coral reef is a structure built up from a definite depth successively and gradually, not by one kind of coral, but by a great variety of kinds, combining together and forming by their joint work a wall, which, from a given depth, may end in reaching the surface of the water. And while it is growing, this wall is all the time changing its builders. It is not one kind that commences and completes the structure to the summit. One kind does a part of the work and then ceases; another kind comes in and continues the work for a while and ceases in its turn; and so on till it is completed.

Here we have a slanting shore. Suppose at six

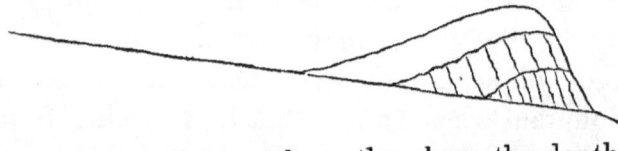

hundred feet distance from the shore the depth is ten or twelve fathoms; it will be a favorable level for the formation of a succession of reefs, for the animals which begin the work live at that depth.

They commence building a wall in that form, — steep towards the ocean, slanting gently towards the shore, rising in the end to the level of the water. The steepness of the outward wall and the gentle sloping towards the land are the result of those fostering influences which accelerate the growth of the reef under conditions which are most favorable to the development of different corals. On the one hand the effect of muddy water, occasioned by storms and tides raising the sand and mud at the shore, is to destroy the corals near the shore and prevent the building of the reef. On the other hand where there is a somewhat steeply slanting shore and the water is pure and plentiful, the conditions are most favorable to the animals. Consequently, on the side towards the sea the favorable conditions are increased, while towards the land they are diminished. The wall therefore towards the sea will be built up almost vertically, and will grow more rapidly than that towards the land. Hence you see that there will be a gentle slope towards the land, as here represented.

But one thing must be remembered. The Radiates which begin the reef after building it up to a certain height necessarily create conditions that are unfavorable to their growth. The condition of the water inside, towards the land, is so altered that the first set of corals can no longer prosper there. The space inside becomes almost an inland pool, even though the water washes over the top of the wall. And now another kind of coral sets in and begins to build. The work goes on, but not so rapidly, perhaps, as before. The first set stops at a certain height; the second set carries it up higher towards the surface. The second set are more hardy, and require less of the immediate action of the sea to sustain their growth. But there are still other kinds which never build the reef itself; namely, those which grow under its shelter. They may be compared to the underbrush of the forest, which does not begin till the forest-trees have reached a certain height. So we have the reef-builders and the underbrush. And then still a third set of reef-builders may come in and bring it up to the level of the water; and after they have grown, the underbrush fills up the bottom towards the land.

Now comes a question which, for a length of time, was one of the most perplexing in the study of these animals. Having ascertained that different portions of the reef, at different depths, are built by different species, and that all these Corals are immovably attached together, the question arises, Whence did these new Corals come which have built up the later portions of the reef?

On examining these animals we find, along the partitions which divide the internal cavity, bunches of eggs. They have been long known as such. But what was not known is the fact that the young which are hatched from these eggs are free and swim in the water. They are little pear-shaped bodies surrounded with innumerable fringes which keep them revolving in the water. They move about at will until they find a proper resting-place, where they fix themselves and grow. Whenever there is a reef which has grown up to the level, say, of six fathoms, where the second set of Corals come in, there will be found these little floating animals, which subsequently attach themselves to the reef at their proper level, and grow. Then another set will come in, in the same way, find their proper resting-place, and so build up the reef.

The succession of these different species of animals is now readily explained. Each one of these little young animals undergoes a transformation from a free swimming body to a Polyp. The method of that transformation has already been explained in a preceding lecture.

The next question is, how long does it take a Coral reef to grow from its base to the level of the sea? They begin, as I have said, at the depth of ten or twelve fathoms, so that the height of one of these reefs is from sixty to seventy-two feet.

Now, here I will make a statement which those who have read the voyages of Captain Cook will perhaps discredit. Captain Cook, or rather Mr. Foster, who was his scientific companion, brought from the depth of two thousand feet in the Pacific Ocean fragments of Coral; and from that time it has been generally, if not universally believed, that Coral may grow at that depth. And yet I say that there is no evidence whatever that Coral reefs grow at the depth of more than twelve fathoms. Now I do not deny the fact that Mr. Foster did bring up Corals from the depth of two thousand feet. But they were dead Corals! Living Coral reefs are never found below twelve fathoms. And since we know that in our own day the Pacific Ocean is subsiding, and even in what direction that subsidence takes place, is there any reason to marvel at the fact of finding remains of Coral reefs at the depth of two thousand feet? They are Corals which have long been dead, and since the period of their death they have subsided with the land to that great depth.

The rate of growth is an important item in the solution of the present problem; and it is the more important that it should be accurately determined, since it must form the basis of the estimate of the age of these reefs. It would be trying your patience for me to attempt to give the full evidence upon this subject. I will mention only a portion of it.

There are on the southern coast of Florida several Coral formations of great magnitude, the foundations of which were, at a date which is recorded in books, laid under the level of the water. From those records I have ascertained that within fourteen years (the period which has elapsed since I made an examination of the foundations of Fort Taylor at Key West, and Fort Jefferson at Tortugas Islands) the addition in the way of a crust of Corals formed upon those new artificial structures does not exceed an inch. Therefore less than a foot would grow in a century. But branching Corals do not occupy the whole of the ground over which they spread, any more than trees. And if I should make allowance for the addition that is due to the accumulated material of the Coral, — as the wood of a forest, for example, if reduced to powder, would add considerably to the thickness of the soil, — it would perhaps reduce my estimate one half or thereabouts. But that there may be no cavil at my data, I will say that an inch in fourteen years, or, to make it easier, a foot in a century, is the amount which the Coral reef is likely to add to the thickness of the soil on which it grows. And by so doing I have certainly overrated it more than twice. How long will it take, at that rate, for a reef sixty feet high to grow? Six thousand years. That, then, is the age we may ascribe to one reef. And if my standard is too large by double, as it probably is, then it would take twelve thousand years. But we will put it at six thousand.

Now let us see how these reefs are arranged around the extremity of Florida. Here is a map of that end of the peninsula. Outside of the main land is a series of islands, known under the name of Keys, the westernmost of which is Key West. Outside of these Keys is a succession of very small islands, very 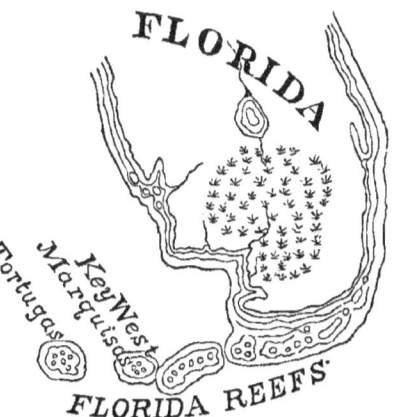 much scattered, but all resting upon what is known as the Florida Reef. They rise just above the summit of that reef, the whole of which is made up of living Corals, — not only the crest, but the sides.

The reef has just grown to such a height that the crest begins to emerge above the surface of the water. This outer row of islands, therefore, must have required six thousand years to form. We will put that down as the first item in our estimate of the age of that region of country.

Now let me give you a few geographical details respecting that country, so that you may more readily take in the facts of history. The reef is separated from the Keys by a shallow channel ranging from two or three fathoms on the right, to six on the left. Then we have a series of Keys separated one from the other by shallow cuts, and from the peninsula by mud flats in which there·is rarely more than a few fathoms of water. Indeed, sometimes large tracts of these mud flats are uncovered at low water. The shore of the peninsula is made up of bluffs rarely more than ten feet in height. The highest peak on the shore is thirteen feet, and the highest of any of these islands not more than that. We have such crests all along the shore. But north of the shore we have what are known as the hunting-grounds of the Indians, — low, flat, marshy grounds, hardly above the level of the sea. Still further in the interior we find what are known as hummocks, — little hilly tracts of land, from five to ten feet above the level of the sea, strangely arranged in a row.

When I examined these Keys under the direction of the Superintendent of the Coast Survey, who always takes a pride in making scientific investigations, I found that they were reef and nothing else, and that they differed in no way from the reef which is growing at present, except that they were cut at their summit as by loose material breaking off from an old reef; in fact, that they were an old reef altogether dead, the materials of which had partly been broken off at the top by the action of the tides and storms.

Now the question is, How could that reef have grown inside of the other, when the evidence is that not even single Corals will grow inside of a reef already grown? On examining and comparing the outer reef with this row of Keys, I very soon came to the conclusion that when these Keys were growing, there was no outer reef at all; that the inner reef had all the conditions favorable to growth which the outer one has now; that it was formed in the same way; that it rose from its foundation to the surface of the water, then died, and then at the distance of some three or five miles, the conditions favorable to the formation of another reef having been laid, the reef which is now nearly completed began to be built.

But there have been soundings made in that reef. It is as thick as the other; it has come up from the same depth, and it must have taken as long to grow as the outer reef. We have therefore evidence here that six thousand years before the outer reef began to grow there was another reef beginning to grow nearer the shore, which must have taken as much time to complete as the outer reef, and which now forms the series of islands called Keys. This is the second item in our calculation.

Upon examining the mud flats inside, and comparing them with the mud flats accumulating outside of the reef, we find that they are exactly the same. In some places the depth is from two to six fathoms, while in others the channel which separates the shore from the Keys has been filled with mud, so as almost to connect them with the main land.

Having satisfied myself of the fact that these Keys were nothing but a Coral reef which had ceased to grow, I became exceedingly interested in ascertaining the nature and character of the main land, and accordingly extended my explorations to the shore. So closely connected are scientific investigations that to carry them out immediately for a present practical purpose is impossible; and the sooner the community understand this, the sooner will they get rid of pretenders and false researches. A careful examination of the geology of the shore led me to the conviction that here again we have nothing but a reef identical in structure with the Keys, just as the Keys are identical with the outer reef. How could that reef have grown with a dry barrier outside of it? It could only have grown when the Keys did not exist, when the most favorable conditions prevailed for a reef to grow along the shore of the continent. And this formation on the shore has been measured, and found to have an average thickness of the Keys; therefore it must have taken as great a length of time to form, that is, six thousand years at least. So then we have a third item of six thousand years to add to our chronology.

But within these shore bluffs and the Indian hunting-grounds are hummocks, and these hummocks — not to extend this demonstration further in detail — are another Coral reef, concentric with that on the shore. They have been built by the same animal and have the same structure. I have collected all the species which are alive at present on the reef, and compared them with those which formed these hummocks, and they are the same. So that, if there is any accuracy in these two leading facts, namely, that the rate of growth is less than a foot in a century, and that the existence of an outside reef precludes the formation of a reef inside, we have the evidence, in the existence of these four concentric reefs, that twenty-four thousand years ago there was a sea washing the place where these hummocks are, and that no reef had then formed beneath them.

And yet this is not all. All these animals are of the same kind as those that live now, and what I have described to you is only a narrow tract of only some fifteen or sixteen miles. More than sixty miles in the interior is Lake Okeechobee, and though I have not myself penetrated as far as that, intelligent officers of the United States army, who travelled over the whole of that country during the Seminole War, have told me that the whole country to Lake Okeechobee is made up of similar hummocks in concentric lines. Now if we take into account the fact that at Tallahassee and Augustine we have the same Coral formation, we have more reason to assume that this has not been growing at another rate than that at which the extremity of the peninsula is growing. Taking, therefore, the distance of sixty miles upon this basis, you see it opens a prospect of chronology for which we are hardly prepared. We shrink even from the evidence that it has required twenty-four thousand years to build this narrow strip of land; how shall we shrink from the assumption that hundreds of thousands of years must have been required to build that prolongation of the peninsula of Florida which is entirely made up of Coral reefs! And yet what is that compared to the age of the world? It is to-day! It is modern time! It is the period which geologists call the present, for it is a period within which the species of animals which now live began to exist on the earth!

I shall show you hereafter that geology has established a chronology which is still less within the limits of our comprehension, demonstrating more conclusively than by anything I have been able to bring before you to-night, that it is indispensably necessary to separate the chronology relating to human events from that which relates to events in the physical world, and that to identify the age of mankind with that of the world, or even of our earth, is to confound things which have hardly any relation to one another.

LECTURE IV.

PHYSICAL HISTORY OF THE EARTH. — MAN THE ULTIMATE OBJECT.

Ladies and Gentlemen: — I attempted in my last lecture to satisfy you that the phenomena of nature, and especially those relating to the existence of animals upon earth, must be measured by a chronology different from that by which human events are recorded. Man has passed through a short history upon this home of his, while we know from observation of animals that their existence must be counted by hundreds of thousands of years, showing that the standard of measurement of their existence is very different from that by which we measure events in which the human family is interested.

There are other phenomena which would afford similar evidences, but which I must pass by as they are not necessary to show that these views are capable of demonstration. I will only allude to one or two facts. The manner in which the Niagara River has worn a channel below the Falls affords undoubted proof of the long period that it has taken the river to produce that result. And we should arrive at similar conclusions if we examined the loose materials which have been accumulating at the mouths of the great rivers of the world. The delta of the Nile, which has been solid land since the dawn of human history, is the work of that river, and shows that it was flowing when there were no men on earth. Our own Mississippi tells the same history. We have no tradition which accounts for the accumulation around those prominent forks of the river which project below New Orleans; and yet what is that space compared with the whole distance from St. Louis to the Gulf of Mexico, all of which has become dry land by the deposit of mud brought down by that river? For there is geological evidence that the Gulf of Mexico once extended even to the upper bend of the Missouri. With such data before us, we are at a loss to appreciate the duration of the periods which are embraced in that modern epoch which, geologically speaking, constitutes the present of our earth; for all these phenomena are contemporary with the animals and plants which now exist. We find in these accumulations of mud the remains of the same plants which grow on the adjacent hills, the same shells, and the bones of the same fishes that now exist, just as we find the same corals in the reefs inside of the Florida coast and those outside.

But there is one feature in the growth of coral reefs to which I have not alluded, and to which I beg leave to call your attention for a moment. Simple as the structure of these animals is, there is yet a difference in their respective standing. Some, by the complication of their structure, rank higher than others, so that there is a possibility of establishing between them a gradation from the lower to the higher; and we shall see hereafter that that same gradation may be traced also in the order in which animals have been introduced upon the surface of the earth. Now this gradation in corals is simply this: Those which are more compact in form are inferior to those which divide or radiate in elegant branches. Thus in the successive complications of structure there is thought manifested of successive improvements. And that same gradation we observe in the various levels at which the corals grow. Those which are found deepest are lowest in gradation or structure, while those which grow at the level of low-water mark are the highest in structure. So completely is nature imbued with this plan, — with the thought of successive gradation, — that even in these walls, built up by corals in the depths of the sea, we read the mind of the Creator, as well as in those higher developments which characterize the structure of animals and assign to each class its respective standing, and also in the order in which they have been introduced upon earth from the earliest periods to the present time.

But the evidence of the long period necessary for the introduction of all these beings would be incomplete, were I not now to give you a short account of the physical history of our earth as far back as it can be traced, calling to assistance the evidence furnished by astronomy on one hand and geology on the other. Within the limits of our own solar system we have a few hints respecting the earlier condition of planetary bodies and the mode of formation of our whole solar system. The existence of Saturn's rings is a phenomenon easily explained by the experiments of Plateau, which show that masses revolving upon a common axis are gradually spread at the equator, and finally divide and form separate bodies. Let me give a brief explanation of his experiments. Plateau, by combining alcohol and water, formed a liquid of a specific gravity equal to oil. Of course oil being poured in would no longer float on the surface, but would form a cluster or globe in the centre and remain balanced. Introducing a rod into the vessel and setting it rotating, the mass of oil was gradually brought to revolve with the rod. As the rotation increased, the mass of oil became projecting at the equator and flattened at the poles. Accelerating the motion still more, the portions at the equator had a tendency to fly off, until at last, by increasing the velocity, the extreme portions separated entirely and formed a ring, like Saturn's; these, after rotating a short time around the mass, were again attracted to the centre. And as this ring lost its rotation, the tendency was to flow together in a spherical form, constituting a body revolving around the central mass. Thus was exemplified, no doubt, the way in which our whole solar system has been formed, if we start with the generally received hypothesis of La Place, that the solar system was once a nebula similar to those which we see in space, and which, by the highest telescopic power, cannot be resolved into distinct bodies. Such a mass set revolving would necessarily assume an ellipsoid form. Portions of the periphery would be thrown 'off first, then others, then others, which would revolve around the central body, so that, finally, a greater or less number of these isolated portions would form planets. Then, if one of these planets should repeat the same phenomena, we would have satellites.

But there is no means of ascertaining what lapse of time it has required for this process, and yet we have to take this part of the history into account in the chronology of the physical world. The earth being once set off from its primary has thenceforth a history of its own, and the question is, what is that history from that time to the present? There remains here a field of investigation to be explored. The history as presented by the astronomer is not yet taken up by the geologist and followed up. There is a middle ground which requires both astronomical and geological knowledge; and owing to the deficiency of our means of investigation, and to the fact that there are no scientific minds prepared for that kind of investigation, we must wait till another generation of scientific men is educated for that purpose.

When geologists take up the subject it is with our earth as a distinct body, already solidified at its surface, presenting two kinds of rocks of different origin. These two classes of rocks we trace everywhere as forming the crest of our globe, namely, first, those rocks in which we see no regular divisions, only unbroken masses; and second, those rocks which are divided into layers, one above the other. These two classes have unquestionably been formed in different ways. Those which occur in layers have been deposited by the agency of water; for everywhere we see that loose materials, when dropped in water, form such layers at the bottom. On the other hand, the other kinds of rocks present such a similarity in their structure to the lavas which flow from volcanoes, that it has been demonstrated that they must have had an igneous origin, — that they were first in a state of fusion, and afterwards became solid. Such are all the granites, the syenites, the porphyries, the serpentines, and the like. Nowhere do they present regular divisions into beds and layers, as we observe among the rocks which consist of grains of sand, or particles of lime or clay.

One of the most direct evidences that the stratified rocks have been deposited by water is derived from the fact that in all of those rocks we find remains of animals and plants in as varied a condition as they are now found in the alluvium along our coast. They form an integral part of the rock itself, and their position in the rock shows that the beds must have been deposited in horizontal layers.

This is an important point, because, if it can be demonstrated, consequences of great magnitude follow with reference to the changes which these

PHYSICAL HISTORY OF THE EARTH. 75

beds have undergone. I will therefore present that evidence so far as it is necessary.

Heavy materials falling to the bottom of the water always fall on their broadest surfaces. Throw an oyster shell into the water and it will reach the bottom on its side. Rarely will it be planted in the mud upon its edge. Now, remains of shells are found in all stratified rocks, and everywhere are they found lying parallel to the surface of stratification. You may remove layer after layer, and find the uniform position of the shell to be with its broad surface parallel to the surface of the rock. And yet we find stratified rocks in very different positions, which leads to the conclusion that whenever they are found in a slanting position they must have assumed it after their formation; and that if they had remained in a plastic state, foreign bodies deposited in them would have resumed a position in accordance with their weight, their greatest surface being parallel to the horizon.

As regards the unstratified rocks we never find in them any traces of organic remains. And upon this broad evidence, which the limits of the present lecture will not allow me to present more fully, it is generally admitted by geologists that these two classes of rocks have been formed in different ways, —that the massive or Plutonic rocks were once in a state of fusion, and that the stratified rocks are the result of the accumulation and deposit of loose materials at the bottom of the water.

Now the point where geology takes up the history of our earth is when the whole of this globe

was in all probability a mass of melted material.
How it came to assume that state is not understood; but the evidence, that the material forming the interior of the earth must be in that condition, is derived from the fact, that, as we penetrate through the stratified layers and reach the rocks below, they present no stratification. Here for instance we have a mountain. The interior is a mass

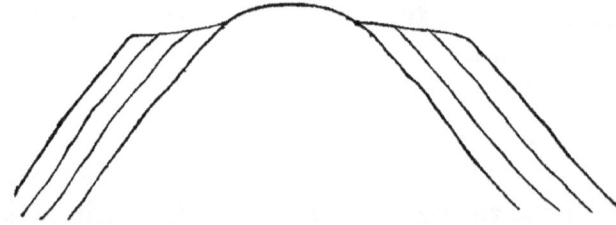

of granite. On the sides are planted beds of stratified rocks, sloping in opposite directions on the two sides. It is evident that the mass of granite is inferior to the beds on the sides, though it rises higher; for these beds can easily be ascertained to correspond to one another, the lowest on one side corresponding to the lowest on the other, and so on through the successive beds to the highest. Such demonstration in mountain regions is very easy. The Jura and the Alps in Switzerland present for geological investigation diagrams from nature on a grand scale in which these facts can be shown just as I present them on the black-board. The nucleus of the mountain is made of masses of unstratified rock, and the slopes are made up of beds slanting in opposite directions; and frequently the lower beds resting upon this mass can be seen to close over the top. It is a matter of demonstration that the want of continuity is the result of breaks which have taken place when these beds were removed from their primitive horizontal position into that which they now occupy, in consequence of an upheaval or pressure from below, causing a protrusion of the masses beneath through the stratified layers. Thus we have mountain chains the centre of which consists of masses upheaved from below, flanked by the superficial strata thus displaced. And when this evidence is connected with the fact that in all these beds oyster-shells and other remains of animals and plants are found in the slanting position assumed by the beds themselves, and that on the opposite sides are found the same kinds of remains in the same relative position and order, the evidence is overwhelming that those beds must have been originally formed in a horizontal position, where they remained until they hardened into rock.

Whatever be the age of these beds, we find everywhere below them unstratified masses, showing that rocks of igneous origin are the foundation crust of our earth.

Let us connect with this another fact; to wit, that these beds must have been deposited one after the other, the lowest first and the uppermost last; and if everywhere we find below the lowest beds unstratified masses, we must come to the conclusion that there must have been a time when there was nothing at all above the unstratified masses, and when the material out of which the unstratified beds were formed was all there was that constituted the solid portion of our earth.

That this material must have been in a state of igneous fluidity is demonstrable by geological evidence. For it is found everywhere that at the point of contact between the unstratified and the stratified masses, the stratified rocks are altered in the same manner that heat would alter them. If you throw limestone in the fire, you make quicklime of it. If you throw sandstone in the furnace, you make coarse glass of it. You alter those substances in the same manner that heat would have altered these stratified beds. Wherever the stratified beds are found in contact with the unstratified rocks, they present alterations which are identical with those produced by heat, thus affording undoubted evidence that the unstratified rocks were once in a state of fusion. And this is further demonstrated by the fact that in most of the great mountain chains cavities are found where these unstratified masses have flowed in, filling them up with solid material. The sides of the cavities are actually soldered together in this way. There is therefore no escape from the conclusion that the stratified rocks were deposited, the lowest first and the uppermost last, and that the unstratified rocks are the oldest.

But even these unstratified rocks are not all of one age; the eruptions of the igneous mass have not all taken place at one time. The largest volcanoes in our day are pouring forth melted material; so there have been such eruptions at all times. The difficulty is in a given case to trace the relative age of the unstratified and the stratified masses.

That geologists are now doing, and they have accomplished it to a great extent.

I will therefore proceed to show you, 1st, In what order the stratified beds have been deposited. 2d, In what order organized beings — animals and plants — have been introduced upon earth. 3d, That in this order of succession we recognize a plan, the ultimate object of which was the introduction of man on earth. For though creation has been carried on so long, it is a matter susceptible of demonstration that at the very outset, when the first animals and plants were called into existence, they were so constructed as to show that they involved the plan of the creation of man.

The unstratified rocks, as I have shown, are the oldest. They contain no traces of the remains of either animals or plants, and therefore furnish evidence that there was a time when the earth was not inhabited; for there are hardly any animals so soft that none of their parts could be preserved. The solid parts of animals, when once deposited in sand or mud and covered, are there preserved and treasured up for all future time in the solid rock that is formed out of the deposit. In exploring the strata of our earth and examining their contents, geologists have become acquainted with the various animals and plants that have inhabited our globe in the early periods; and their number is so great that the conclusion is inevitable that at all times, since the stratified rocks have been forming, the earth has teemed with inhabitants as various and diversified as they are now. Within the limits of this state there are beds of rock so full of remains of animals and plants that the mass of strata consists of almost nothing else. Indeed, along our seashores we do not find such quantities of dead shells as we find in some of the limestone rocks in the western part of the State of New York. And yet these rocks are among the oldest of the stratified beds on the surface of the earth. But below these are found masses of rock in which no trace of organic remains are found.

Having now the first appearance of these remains upon earth in the earlier stratified rocks, we can begin to trace the gradual succession of the stratification, and so form a complete history of the development and progress of life from the beginning to the present day. We may consider these beds or layers as the leaves of a great book, in which is recorded the history of all the various plants and animals that have existed since creation, and the order of their succession. The ability to read this book is not very difficult. It is only necessary to study living animals and plants to such an extent as to be able to recognize them when they are not quite perfect. Now as a skilful surgeon is able to recognize every bone in the human body when presented singly and separately, so it is possible for the naturalist to recognize each tooth of any common animal, and to assign it its proper place in the mouth. So different are the teeth of the various animals, — the horse, the cow, the dog, the cat, — that he who has studied them minutely can recognize at once to which animal any single tooth belongs. And this study may be extended so far that it will not be difficult to distinguish any separate bone of an animal and tell to what species it belongs. So also may the different scales of fishes be recognized and distinguished even when separately presented. In fact there are those who can not only recognize any fish from a single scale, but can tell to what place on the body it belongs. And so with a dead shell; a fragment will indicate the particular kind of animal that it is part of. Now when you have acquired sufficient familiarity with the anatomy of the various living animals, you may apply that knowledge to animals which no longer live on earth, — to remains of those which have existed in former ages; and when even fragments of those animals are found in the rocks, a comparison between them and animals now living will be sufficient to ascertain how far they are related to one another, and whether they belong to the same or to different species. To such an extent has this study been carried that geologists are able to refer these remains at once to their different natural groups. And the result of their investigation is this, that all the animals which live now upon the earth, to whatever class they may belong, differ from all the animals which have existed before; and those animals which have existed in the earlier periods are not the same at different depths of the earth's crust, but each successive bed contains remains entirely different from those above it. So that we may consider the whole stratified crust of our earth as an immense cemetery made up of large leaves or sheets of rocks, within which are buried all the varieties of inhabitants which have lived on the globe, and to such an extent preserved that their true character can be identified. And in this work of restoration of the earlier inhabitants of our globe naturalists have already proceeded so far that it is no exaggeration to say that the remains of some of the animals of those periods are better known to us now than some that exist at the present time which have not been brought within the range of civilization; as for instance, certain animals inhabiting the interior of Australia, or of Africa, or even our own continent. This is owing to the fact that fossil remains are in certain localities to be found in immense quantities, so that we may make collections as complete as we can of butterflies in summer.

Knowing now the level at which the first animals and plants were called into existence, as we proceed with the examination of the successive beds of rocks, we find that there are certain breaks or interruptions of the order of succession, indicating great commotions on the globe. It is well known, that, at the beginning of the Christian era, Monte Nuovo, in the Bay of Naples, rose above the level of the sea about four hundred feet, where it has remained ever since. In that upheaval the water was so shaken that fishes died in immense numbers; and for a great distance in the surrounding region the bottom of the sea was elevated, causing innumerable marine animals to perish. Before that eruption everything was going on quietly in those waters: the sea was beating the shore, animals were dying in the course of nature and sinking in the mud in regular order. But when this mountain rose those beds were disturbed in the same manner, probably, as in former ages, though on not so grand a scale, when immense portions of the earth's strata were upheaved and erected into mountain chains. It is those gigantic upheavals, no doubt, which have changed the configuration of the earth's surface, and produced those interruptions in the regular order of succession which we observe in the strata of rocks.

Now if we take advantage of these breaks to separate and classify the beds which form the bulk of the earth's crust, we find that they may be divided into some twelve different systems, which geologists have named as follows: —

The lowest, or those which contain no animal remains, are called Azoic. Upon them are deposited the lowest formations which contain animal remains. These are subdivided, but the name of the general system is Silurian. Next we have the Devonian; then the Carboniferous, the Permian, the Triassic, the Jurasic, the Cretaceous, the Eocene, Miocene, and Pliocene; and, lastly, those beds of Alluvium which form the present system in which deposits are still going on.

Now these different sets of beds mark as many different epochs in the history of our globe. Each is characterized by peculiar animal and vegetable remains; in none of them do we find identical fossils.

There has been recently entertained by some a favorite idea that all animals began by a few representatives, which have been gradually improving and changing until all the diversity which now exists was produced. I have already alluded to the domestic animals as furnishing no evidence whatever of any such theory. And I introduced the subject of corals with a view of showing that we had positive evidence that, outside of the direct, fostering care of man, animals do not change during immensely long periods. And now geology furnishes us the most direct evidence upon the same point. It shows that there has been no such gradual transformation, but on the contrary that there has been the same diversity, which we observe now, in all times.

We find that all the different types of animals existed in the most ancient times. Representatives of the four great divisions — Radiates, Mollusks, Articulates, and Vertebrates — have always existed side by side. These, therefore, could not have been derived from one another, for contemporaries cannot be each other's descendants. On the lowest layer on which remains of animals have been found, we discover various kinds of Radiates, Mollusks, and Articulates. And though we do not find Mammals, Birds, or Reptiles, yet we do find Fishes there also, so that at least the type of the Vertebrates is represented, showing that there was an intention of introducing these classes in such a manner as to indicate their relation to the highest being on earth, — Man.

Again, while all these types have been represented from the beginning, at each successive period they have been represented by different kinds. The Radiates that are found in the lowest beds are not the same as those found in the Devonian system, while those found in the Carboniferous system are still different, and so of the Permian, the Triassic, the Jurassic, the Cretaceous, the Eocene, the Miocene, and the Pliocene. They differ, as found in all these systems, from one another, as well as from those which are now living. And this is also true of the Mollusks, the Articulates, and the Vertebrates; and though of the latter we have only Fishes first, at a later period we find Reptiles, still later Birds, still later Mammals, and during the last period Man.

The character of this succession I shall explain in a future lecture. I will only lay before you now the evidence that all the principal divisions of the animal kingdom have had a common start; that they originated at the same time, and that therefore whatever ideas are involved in the plan of creation, the whole plan dates back from the beginning, and involves all the combinations which are presented in time. The earliest of them show that they are intimately linked to all those of a later period; but to that I shall devote my next lecture.

But there is one point I want to mention to-night, in order that I may not have to interrupt the demonstration in the next lecture; it is the nature of the evidence that these animals have discontinued their existence; that there are breaks or interruptions at which the inhabitants of the earlier periods ceased to exist, and were replaced by the representatives of another period. Whenever we examine the slopes of mountains we find that the beds which have been raised by upheavals do not present everywhere the same relation to one another. In the Jura, for instance, the lowest

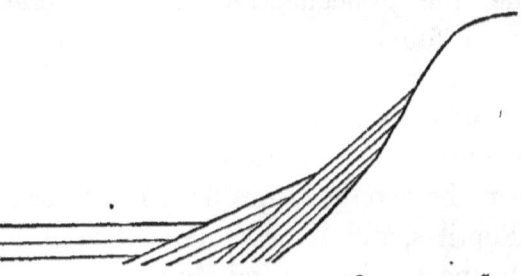

beds have a certain slant; upon these are deposited other beds that slant at a different angle, and upon these we have still other beds which are perfectly horizontal. Here we have the evidence of the upheaval of that mountain at successive times. It is perfectly plain that when these lower beds were raised to their present position, the upper beds did not exist. We must assume that the second set of strata were deposited after the first were upheaved, and were deposited in a horizontal position against them, and that after they had lain so for an unknown length of time, they were themselves raised into their present position, and the uppermost beds were deposited upon them afterwards.

And now if we can ascertain that the first set of beds is Jurassic, the second Cretaceous, and the third Miocene, we know that that mountain chain was raised earlier than the Cretaceous period, and that before the Miocene period the Jura underwent a second elevation. It is by such facts that geologists have been able to determine the relative age of all the mountain chains on the globe. And the conclusion of this investigation is this, — that the higher mountains are younger than the lower; that all the highest mountain chains have been upheaved in the most recent period, prior, of course, to the present epoch, for the phenomenon is one that has never been witnessed by man.

If we examine the Alps we find that there have been upheavals not only of Jurassic and Cretaceous beds, but also of Miocene, showing that the Alps were raised subsequent to the Jura. The Jura, therefore, is the older mountain chain. And we find similar facts in regard to mountains in other parts of the globe, all proving that mountain chains are older in proportion as they are lower, or younger in proportion as they are higher.

And this you will see is a natural conclusion, and one which we should have expected if we had reasoned on a sound basis. The lofty mountains were once considered as the backbone of the earth, around which the waves of the ocean had been heaping up loose materials. But if the internal structure of the earth had been known as it now is, it would at once have appeared, that, before any of the stratified beds had been deposited, the crust of the earth being thinner than it now is, whenever an upheaval took place, there being no great resistance by the earth's crust, only small hills would be formed. It is only where the resistance is great that the result is gigantic. When the crust of the earth became thicker, the internal movements of the melted material became active for a long time before they could overcome this resistance; but when at length they did overcome it, the result was the upheaval of gigantic mountains. So we find all the geological facts to be in accordance with this result. Let us take for example the Rocky Mountains of this continent, and compare them with the Apalachian chain along the Atlantic seaboard. Here we find the sandstone of the Connecticut Valley is slanting everywhere, while the Cretaceous beds, and more recent sand and loam deposits which form the low lands along the shore, are horizontal. Go to the region of the Rocky Mountains, and there you see even the most recent deposits to be found on this continent following the mountain slopes, showing that those mountains are younger than those on the seashore. Thus it also appears that the oldest chain on this continent is that which follows the great Canada lakes, for there we have only the very oldest beds raised so as to follow the undulations of the land, while even the oldest fossil-bearing rocks are at the foot of the hills and do not rise to their summit.

Now, interruptions resulting from the successive depositions of these beds in consequence of such upheavals, mark the limits of these great successive geological periods. During the intervals between those great commotions, the earth has remained quiet as it is now; animals and plants have gone on multiplying, and their remains have become buried in the strata, until another commotion produced another change, another configuration, another disturbance of land and sea, and all the concomitant results. And in each of these sets of beds, as I have stated, we find peculiar combinations of animals, but through all ages consisting of representatives of all the great divisions of the animal kingdom, — at first Polyps, Acalephs, and Echinoderms, associated with Bivalves, Univalves, and Chambered shells, also with Worms, Crustacea, and Fishes, then very soon with Insects, Reptiles, Birds, and Mammals, and at last with Man. And now the order of succession in detail I shall present to you in my next lecture.

LECTURE V.

TRIPLE COINCIDENCE IN THE SUCCESSION, GRADATION, AND GROWTH OF ANIMALS.

Ladies and Gentlemen: — Thus far the train of my argument has been mainly to show that there is order in nature; that the animal kingdom especially has been constructed upon a plan which presupposes the existence of an intelligent being as its Author. But there is one phase of this question far more important in a moral point of view than any I have presented, and to the consideration of which I propose to devote this and the closing lecture. It is that phase which involves the question of the existence of Providence in nature, or in other words, the recognition, on scientific grounds, of the working of a Providence in the world. If nature, as we see it manifested in the facts I have presented, is the result of the working of mind, of intelligence, the result may have been accomplished by one of two methods, namely, 1st, By established laws; or 2d, by direct action. An intelligent Creator may have devised laws as the means by which these results are indirectly obtained; or they may be the direct or immediate work of his hand. A comparison of the works of nature with those of art will at once illustrate my meaning. We recognize intelligence in the construction of a machine because we know that it could not operate in the manner it does were it not the device of an intelligent artisan. But then the work that the machine does is not intellectual work; it is work delegated to it by intelligence, and from that time intelligence has nothing more to do; the machine does the work. Now the question with reference to the existence of living beings, whether they are the products or results of laws working in nature, established by the Almighty, or whether they are the work of the Creator directly, — this is the point I propose to examine on the basis of scientific facts; not on the moral ground upon which we trust in Divine Providence, but upon scientific evidence, for science must deal with facts on its own ground, without reference to preconceived opinions or convictions, and we should welcome what science has to say upon the subject of an overruling Providence.

I have already adverted in my last lecture to a series of facts bearing upon this question. I showed that there had been interruptions in the sequences of organized beings which have existed upon our globe; that the first set of animals had gone on multiplying up to a certain period, or to a certain level, beginning at the lowest formation, and then disappeared to make room for another set of animals, and so in their turn each set of new-comers had vanished to give place to others. But not in the way in which one generation makes room for another. At each period there have existed different kinds of living beings with their successive generations, which, having had a certain duration, have given place to other kinds with their successive generations. So that the earth has been again and again inhabited by different successive generations of different kinds of animals, with interruptions between them, indicating great disturbances in the natural course of events and extensive changes in the prevailing conditions through which the earth has passed, accompanied by successive renewals of its inhabitants.

If we were to credit a certain theory which is very well received at this time, which has lately been propounded by some very learned, but, I venture to say, rather fanciful scientific men, it would appear that in the beginning animals were few in number, and that as they became more and more numerous they became more and more different from one another, as if all the diversity which exists on earth at the present moment had grown out of a comparatively simple and small beginning. This is an impression which prevails so generally that before I take another step in my demonstration I will endeavor to show the fallacy of it.

From the position of the lower strata of the crust of the earth it is apparent that they are not so easily accessible as the superficial ones. The uppermost sheet in a pile is that which we get at most easily, and with greater and greater difficulty do we reach those at the bottom. Now our information respecting the inhabitants which have existed on the surface of our globe at different times is entirely derived from the remains of animals buried in these beds. We know their number and characteristics only in proportion as we can exhume them, and this process is difficult in proportion as they are buried deeper. So you see at once that we are likely to know less of the animals that existed in the most ancient times, even though they may have been as numerous as in later periods.

Then there is another difficulty which has been entirely overlooked. The animals which live at the present time we meet with everywhere. We are impressed with the extraordinary abundance of living beings. The forest teems with trees, the meadow with grass and non-herbaceous plants, and there are living animals in profusion all around us. But fossils, or the remains of animals of past ages, are few. The moment we seek for the inhabitants of the earlier periods, we are restricted in our researches. We are quite familiar with the living animals of Africa and Australia, but what information have we respecting the remains of animals buried in the strata of those continents? Not only Africa and Australia, but the greater part of Asia and South America are sealed books in that respect. It is chiefly in certain parts of Europe and North America that we are enabled to compare the remains of past ages with the animals now living on the whole surface of the globe. In order to make a fair comparison, we should institute it between tracts of the earth's surface of equal extent to those in which we have surveyed animals of past ages in this fossil condition. Let us do so. Let us take as the measure of the variety which has existed in past ages, such tracts of the earth's crust as are accessible to us, in which we can collect the fossils and compare them with the living animals.

The number of species of fishes which inhabit the Mediterranean is only a few hundred; those that inhabit the German Ocean only about one hundred and eighty or two hundred; those on the Atlantic coast of France not more than two hundred and fifty; and yet the sum total of the different kinds of fish known in all parts of the world is nearly ten thousand. If we were to compare the fossil fishes found thus far in the strata of the globe with those of the whole world as they now exist, we should make the same mistake as in estimating the inhabitants of one region as those of the whole world.

The fossil fishes which have been found, and which I have had an opportunity of examining in certain circumscribed regions, form a very favorable basis for comparison and estimate. At Mount Vulcan, near Verona, is a celebrated quarry, not many miles in extent, from which alone have been taken over one hundred different kinds of fossil fishes. The Adriatic in its whole extent does not furnish as many different species as are found in this quarry. I have examined the fossil fishes of the neighborhood of Riga on the Baltic, and they are more numerous than the present living species of the Baltic and German Ocean.

Here, then, we have direct evidence that in former periods, within similar areas, there was as great a diversity of animals as now exists.

But not to confine the comparison to fishes, let us take shells. The number of shells of different kinds in the Mediterranean does not amount to five hundred; in the Red Sea there are not more, nor in any region of the earth of one hundred square miles can there be found more than that number. And yet the whole number of species known at this moment exceeds fifteen thousand. Now, in one single region in the neighborhood of Paris, a single naturalist, Lamarck, after ten years' research, described several hundred different kinds of fossil shells, all of which, of course, are different from any that now exist. The geological survey of the State of New York ordered by the legislature has disclosed in each of the successive sets of beds within the area of this State as numerous a variety of shells as the sum total of all the species now living along the whole Atlantic coast of this continent. What better evidence do we want that at all times the world has been inhabited by as great a diversity of animals as exists now, and that at each period they have been different from those of every other period?

This is a very important fact, because it is a most powerful blow at that theory which would make us believe that all the animals have been derived from a few original beings, which have become diversified and varied in course of time.

I will now proceed to enumerate the animals of the different periods, and then to illustrate their character. I will represent the different systems of rocks in their order by a diagram, dividing them by horizontal lines; and I will indicate by vertical lines the divisions of the animal kingdom.

	Primary.				Secondary.				Tertiary.				
	Taconic.	Cambrian.	Silurian.	Devonian.	Carboniferous.	Permian.	Triassic.	Jurassic.	Cretaceous.	Eocene.	Miocene.	Pliocene.	Present.	
Polyps.														Radiates.
Acalephs.														
Echinoderms.														
Acephala.														Mollusks.
Gasteropoda.														
Cephalopoda.														
Worms.														Articulates.
Crustacea.														
Insects.														
Fishes.														Vertebrates.
Reptiles.														
Birds.														
Marsupials.														
Mammals.														
Man.														

Now, upon examining the lowest of these beds in which remains of animals are found, we find at once representatives of Radiates, Mollusks, Articulates, and Vertebrates. And not only the representatives of those four great primary divisions of the animal kingdom, but all the different classes of the first two of those great divisions. Of the Radiates we have Polyps, Acalephs, and Echinoderms from the beginning. Of the Mollusks, we have Acephala, Gasteropoda, and Cephalopoda from the beginning. Of the Articulates, we have Worms and Crustacea in the lowest strata, but no Insects. The latter only begin to appear in the Carboniferous period. Of the Vertebrates, the only representatives we have in the lowest strata are Fishes. Reptiles are only to be found as we reach the Carboniferous period. Birds not earlier than the Triassic, (if they are found there at all,) and Mammals at á still later epoch, while no trace of Man is found until we reach the present period. Thus we have nine classes of the animal kingdom existing from the first dawn of life; then by a singular coincidence Insects and Reptiles are introduced together at a later period; then at a still later period we find Birds, still later, Mammals, and lastly, Man.

Let us now proceed to examine the character of this succession. So striking are the differences between the various classes of animals at different periods that geologists are able to recognize almost at first sight the coral of the Carboniferous period as readily as a shell from the Gulf of Mexico, the Red Sea, or Australia. Animals are distributed on the surface of the globe according to definite laws, and with remarkable regularity. There is no disorder in their distribution, only it requires long study before we can grasp their diversity to such an extent as to be able to understand how they are combined on the surface of the globe. You know very well that if you wish to see a Palm you must travel to the south, and if you would see a Pine forest without any deciduous leaves you must go very far north, — for in our own latitude we have a mixture of permanent with deciduous leaves. Now this study may be carried so far as to show that each species of plants and animals has its definite home, and the power with which animals are endowed of locomotion is used by them not to wander at random over the surface of the globe, but to roam within genial regions.

Now in the order of succession we find something quite similar. It does not require many years' study of fossils to recognize at a glance a chambered shell as belonging to this, that, or the other geological period. Their characteristics are as distinct and as easily recognized as the pine is distinguished from the palm. And with these differences is very soon associated the idea of an earlier or later existence, just as you associate the idea of a warm climate with the palm and a cold climate with the pine. And not only is there order in this succession, but there is an order which shows at all times consecutive thought, which at the outset perceives the end. This is something which is never put by mind into machinery; it is something that the architect retains to himself only while he is superintending the work. In the combinations which are observable among the representatives of the earlier period we can discover that relation to one another which at the very beginning implies that the end is perceived.

In presenting this argument, in the brief limits that are allowed me, I will select for illustration the class of Echinoderms. Of these there are a great variety of representatives living all over the surface of the globe, and they differ among themselves in the complication of their structure to such an extent that they have been classified in several orders. One of those orders embraces Star-fishes, which have a star-shaped form; another embraces Sea-urchins, which have more or less a hemispherical form; and a third order is called *Bêche-de-mer*, having a cylindrical form. There is common to all these animals, whether star-shaped, spheroidal, or cylindrical, a mouth in the centre from which all parts radiate in every direction; for the ribs which extend from pole to pole in the cylindrical animal correspond to the rows which we see on the hemispherical body of the Sea-urchin and to the rays of the Star-fish.

But this is not all; the upper and lower side of the body are made up of different elements. There are Star-fishes in which those elements are closely packed together and form a sort of calyx, which may be attached to the ground by a solid stem, while from the upper portion only branches radiate. Such animals are attached to the ground, and incapable of locomotion. They have received the name of Crinoids, while Star-fishes proper are called

Asterians, and then there is another group called Ophurians, — animals with a sort of disc in the centre from which arms radiate very abruptly.

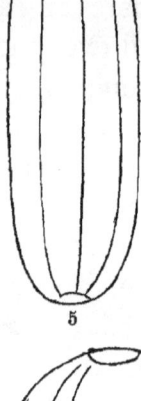

5

Now you perceive at once that there is a gradation in these animals. First and lowest we have Star-fishes with a stem attached to the ground; next those which radiate equally in every direction; then animals with a central disc and radiating arms; then those which have a hemispherical form with certain anatomical complications, which, without entering into the details, require us to assign them to the fourth position; and lastly, we have the cylindrical-formed animal which is the highest in structure, being the most complicated of all.

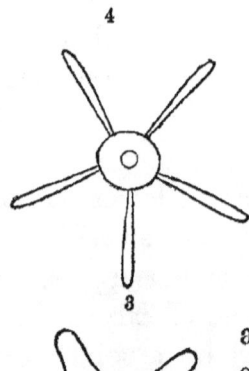

As to the frequency of the types of these different animals, we find only one single kind of the description of No. 1 now known to exist; that is found in the West Indies, about Porto Rico, and is called Pentacrinus; while of the kind of No. 2 there are more, of No. 3 still more, and so on. Those occupying a higher position are very numerous, while those occupying the lowest places are exceedingly few.

But let us inquire in regard to

the Sea-urchins of past ages and those of the present. We need only go to Lockport or any part of Western New York to find in the lowest beds an innumerable quantity of Echinoderms. But they are all of the kind of No. 1, — nothing but Crinoids.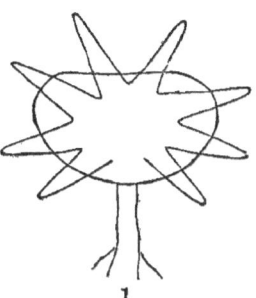
In one locality at Lockport there are as many different kinds of Echinoderms of that family as exist of all living species along the whole coast of the United States. Beginning, then, at the lowest of these beds, we follow up through the limestone of Pennsylvania, which is superior in position to that of New York, and so on through the carboniferous rock. Here we find quite as great a variety of Star-fishes as in the strata of New York, but of different kinds, not a single one like those of the lower strata. Then a little higher up we find genuine Star-fishes, and still higher genuine Sea-urchins. Then as we rise higher these animals are very numerous, but we have not a single one of No. 5 in the early geological periods; they all live at the present time.

So what here strikes the observer is the fact that the order in which these animals have been introduced on earth in the course of time, from the most ancient period up to the present time, presents a similar series to that which we observe in the gradation of the structure of the present living species of the same animals. We have two series which coincide in their result: one an order of succession in time, and the other an order of gradation of structure. It is a coincidence of result obtained by different methods or different ideas.

But this is not all. Let us examine the manner in which these animals grow. The little Star-fish, when it is forming within the egg and when it is hatched, is not the same free animal that it is in the adult state. It is a little being attached to a stem or prong, with branches above, forming a kind of cup. It resembles those Echinoderms first born on the surface of the earth, the type of which has become extinct with the single exception of that one which lives in the Gulf of Mexico. After having lived for some time in that form, it casts off the stem; just as we find in the course of time a period when Star-fishes with stems no longer exist in great numbers, but are succeeded by those that have no stem. So this little animal casts off its stem, becomes free, and assumes the form of its parent.

Now what is there to bring about this coincidence if it is not the mind that has devised the order in which animals should appear on earth, — the mind that has assigned to the lowest in structure the same degree of complication that was given to the oldest in the order of time, — the mind that has established the order of growth of the young animal after the same pattern and upon the same idea that is presented in the order of time and in the gradation of structure?

We have here, then, three different ideas in no

way necessarily connected with one another: 1. The plan upon which animals shall vary in their structure in course of time. 2. The order of gradation of structure of living beings. 3. The order of the growth of the young animal from the egg. And yet in their results those three ideas are the same.

Here, then, we have the work of mind, but not of a mind which acts by necessity, but with the freedom of omnipotence. We have it here directly, and we can demonstrate it the more fully as we trace the facts thus presented more in detail.

It is not in the class to which I have adverted alone that we find these results, but in every class, so far as our researches have gone. Unfortunately we are not so well acquainted with the representatives of every class in past ages as to be able to trace them through the different periods. It is not every class that we can arrange according to complication of structure with unerring certainty. It is not every class in which we have traced the growth of the young from its first formation in the egg through the different stages of its development. But in order to satisfy you that this triple coincidence is not an accidental thing, I will take another class of animals.

The class of Crustacea, among Articulates, embraces Crabs, Lobsters, Shrimps, and the like. The great difference between Crabs and Lobsters consists in this: that while Crabs have a short tail, which is bent under the body and almost concealed, in Lobsters the tail is nearly as bulky as the anterior part. The Crab and the Lobster have similar limbs placed in the same position, the same jaws performing the same functions, and the same feelers projecting from the head. But there is a third group of Crustacea, known as Sand-flies or Shrimps, which have very minute limbs, like hooks; otherwise their main features are the same. Now it requires not much insight into their structure to perceive that these are the lowest, while Lobsters occupy the middle position, and Crabs the highest. In Crabs there is a concentration of parts. The longitudinal nervous swellings become united in the Crab so as to form a centre of sensation more dense than in any other of the Crustacea. Therefore we place Crabs the highest in structure.

Now let us take the Crustacea of past ages. At Trenton, N. Y., are found immense quantities of Trilobites, and any one who has ever seen one of them cannot have failed to notice that they have two distinct regions, one in front and one behind, each different from the middle region, which is

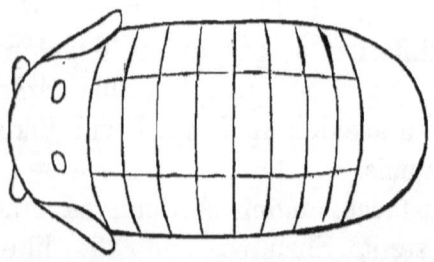

divided by a number of rings transversely, and then by a few longitudinal marks. These Trilobites are found in the lowest fossiliferous strata. There may be such differences as a prong on the side of the shield or a greater prominence given to the eye; but all these are subordinate differences. The prominent characteristic is a body in which the head and tail are hardly distinct, being of the same width as the rest of the body, with uniform transverse divisions of the whole animal, such as we find in the Shrimps of the present day.

Now what do we find in the middle geological ages, — in the Carboniferous, the Permian, and especially the Triassic period? Nothing but Lobster-like Crustacea. In the lithographic quarries of Bavaria they are as numerous as Trilobites at Trenton. But Trenton belongs to the period of the earliest Crustacea, while the quarries of Bavaria belong to the middle geological period. But when you come to Mount Vulcan, to which I have alluded, or to the sandstone which forms the greater part of Switzerland, there you find Crabs, and scarcely anything else. Lobsters are less numerous, and of Trilobites there are none.

How many kinds of Lobsters have you on the coast of the United States? Only a few. Crabs, however, are innumerable. They are the last in order, and are now the dominant tribes on the earth. They stand highest in order of time, Lobsters next, and Shrimps lowest.

Let us see how they grow, and take the young of the Crab; when examined in the egg, long before it is hatched, you find a little oval disc which when it first shows signs of distinct parts presents a shield in front and behind and transverse ribs just like a Trilobite, and so striking is the resemblance that it is not too much to say that the figure of the various kinds of Crustacea in the egg is a diminutive diagram of the Trilobite of the earlier ages.

But when the little Trilobite-like embryo is hatched it does not come out a Crab, but a long-bodied animal, with prominent feelers, and a long tail, just like the Lobster.

So here again we have the same thought manifested as among Echinoderms. The pattern of growth, the order of introduction in time, and the gradation of structure, are coincident. If we follow the course of rocks from the oldest to the present time, we trace the same idea. And if naturalists would study the embryology of Crabs, they may do it as well by resorting to the quarries where the oldest Crustacea are found, as to the eggs of the living animal.

But there is something somewhat different in one of the types of animals to which I will now call your attention. While examining the relation of Polyps, Acalephs, and Echinoderms, we cannot fail to perceive that these three classes are not absolutely higher, one than the other, but that they stand relatively about in this position. But among Vertebrates there is something quite striking. The Fish is unquestionably lower than the Reptile; the Reptile is superior in every respect to the Fish, the Bird is in every respect superior to the Reptile, and among Mammals there are none which we should feel inclined to place below Birds. This gradation we see at once, upon examination of their structure, is a marked feature among them. In the circulation of their blood we find a difference. It is simple among Fishes. Their mode of breathing is through gills; their blood is cold; they lay a large number of eggs, with a very few exceptions taking no care of them whatever. Then we have the class of Reptiles, in which the circulation is more complicated, whose mode of respiration is aërial, and though they lay eggs, those eggs are fewer in number, and there is a more close relation of parent and offspring than among Fishes. Coming to Birds, we have warm blood, a more complicated circulation, fewer eggs, and though in some cases the young when hatched are sufficiently developed to take care of themselves, as among hens and ducks, there are others in which the young are so imperfectly developed that they require the nursing care of the parent. Then, as we come to Mammals, we find a new feature introduced, — the dependence of the young upon the mother, the nourishing of the young by the mother from her own body. And this dependence is proportioned to the standing of the young. There is not so helpless a being born as the human infant, and yet he occupies the highest position according to his organization.

So these four classes are so linked together that from the Fish to Man we have an unbroken succession. The plan of Man's organization begins with the Fish. And we can trace it through the successive geological formations in the same way. In the lowest fossiliferous strata we find Fishes, subsequently we find Reptiles, then Birds, then Mammals, and lastly Man. So here in the order of succession we have a coincidence with their gradation according to structure. And let us see if this coincidence does not exist in their mode of development. Take the egg of a Bird, and examine the growth of the young animal. At first it has all the features of a Fish; the structure coincides very closely. So here again we have the same thought in the mode of development.

Is it, then, too much to say, that, when the first Vertebrate was called into existence, in the shape of a Fish, it was part of the plan of that framework into which its life was moulded, that it should end with Man, the last and highest in the order of succession? We find evidence of this fact in the comparison of the attitude in which he stands with that of the Fish, and also in the comparison of the brain.

Let us examine these relations for a moment. In the Fish the brain is only a slight swelling, scarcely raised above the level of the spinal marrow, which extends through the whole backbone, and the posterior division of the brain is the highest. In the Reptile the posterior, middle, and anterior portions are of the same height, and the whole brain is slightly raised above the level of the spine, for the Tortoise, Lizard, and Snake all raise their heads. In Birds we find the anterior portion the largest, and the posterior portion the smallest, and we have a slanting position of the spine. In quadrupeds we have still further progress. Coming to the noble form of Man we find the brain so organized that the anterior portion covers and protects all the rest so completely that nothing is seen outside, and the brain stands vertically poised on the summit of the backbone. Beyond this there is no further progress, showing that man has reached the highest development of the plan upon which his structure was laid.

But trace this progress also in another aspect. The Fish swims horizontally. His body is all one cylindrical mass, the head does not rise above the rest of the body, and there is no contraction behind it to mark a neck. The Reptile has a slight contraction behind the head; for even the Serpent is not so uniform that you cannot perceive where the body ends and the neck begins, even though there are no limbs to raise that body. But the next step is in the Lizard, where rudimentary legs appear, sometimes capable of raising the body slightly; but even yet, like the Snake, the animal moves mainly by means of the undulation of the backbone instead of its limbs. Then we come to Birds, in which the tendency is to an upright position; the Bird stands on its hind limbs, but yet it has not entirely reached that position. It requires one further step, by which one pair of limbs alone are made to perform the function of locomotion, while the other pair become subservient to the mind. The hand of Man is no longer an organ of locomotion, and it is no longer a paw; it is the organ with which we express our deepest feeling; it is the limb with which we grasp our fellow-being in cordial recognition. The brain of Man occupies not merely the foremost, but the uppermost position. It is not merely forward, but upward: forward in the direction of all progress in intellectual culture; upward in the direction of all moral excellency: forward and upward towards that Mind according to whose image man is made.

LECTURE VI.

EVIDENCE OF AN INTELLIGENT AND CONSTANTLY CREATIVE
MIND IN THE PLANS AND VARIATIONS OF STRUCTURE.

Ladies and Gentlemen: — In presenting in my last lecture the order of succession of animals in past ages, my object was chiefly to show that there exists such a connection between them as bespeaks thought, plan, and deliberation, and that in their combination at different periods is clearly seen the intervention of an intelligent Creator. I propose to-night to complete the argument by showing that the nature of the intelligences is the same whether it be human or divine, finite or infinite. In order to bring this argument near to the comprehension of those who have not studied natural history closely, I will select for this demonstration a class of animals in which the complication of structure is not great.

Nowhere is the evidence of plan more plain, without the deeper study of anatomy, than in the radiated animals. As the basis of their structure we may take the sphere, in which all the points are equally distant from the centre. For though the plan of radiation as exhibited in these animals is not a mathematical sphere, it is nevertheless an organic sphere; it is a sphere loaded with life; it is a sphere in which there is as much differentiation as can well be introduced upon the idea of radiation. And the first difference which we notice is this: that instead of the outer structure bearing a relation to a central point, it bears a relation to an axis which extends from pole to pole. The living sphere, as shown in the radiate animal, comes nearer to a revolving sphere which has an axis of revolution, and which, in consequence of that movement, has two poles and an equator. This axis is surrounded by parts which are identical in their nature, and which bear among themselves the same relation to one another and to the central axis. But the two poles are essentially different.

If we take the problem of radiation in a mathematical point of view, if we present to the mathematician the question involved in the plan of radiation, it will be this: How to execute, with the elements given, — with a vertical axis, around which are arranged parts of equal value, — all the possible variations involved in that plan. This question is not a mere fiction. I have presented it to one of our great mathematicians. I requested him to solve the problem, how to devise structures variously executed, the elements being given, without introducing any new elements. His answer was readily given, and it was this: That the simplest way would be to represent the whole sphere as a series of wedges placed side by side with one another.

And to make this demonstration as clear as I

MIND IN THE PLANS OF STRUCTURE. 113

can, I will take for illustration the melon, the ribs on the outside of which will give the idea of wedges combining to form a spheroidal body. The orange which I hold in my hand would give us the same idea, with one additional element which I will consider presently. Let me take first the inside of the orange. You all know it is divided into a number of parts which are all equal. They are what mathematicians call spherical wedges, the edges of which correspond to the axis, the spherical surfaces of which are segments of a sphere, and the sides of which are the surfaces dividing those segments one from the other.

Now in executing any structure upon the idea of radiation, the simplest way would be to bring together around the axis a number of these spherical wedges until the whole space is occupied by them. There would be a larger or smaller number, according to the angle of the sides of the wedges. If they are thin there will be many, if thick there will be few. So we could introduce at once a variety among them by changing the numbers and perhaps the relative dimensions of them, increasing the thickness of the partition, and modifying the surface which encloses them all.

The substance of the orange is surrounded by a bark. That bark is not however primitively so distinct from the pulp as when it is ripe. When forming, the orange is composed of a number of spherical wedges not yet separate from the bark, which has as yet no special consistency or color. The bark or rind is the result of growth.

These modifications in structure can be further extended by changing the thickness of the wedge near the equator. Also by having each of the wedges hollow, and surrounded by thin walls. Then by making the walls thicker and reducing the cavities. Again, by isolating the surrounding elements, freeing the cavities in the interior and giving them distinct walls; for this is complicating their structure in such manner that they would form independent orders. These are what mathematicians, conversant with all the powers of mathematical combinations, present as the various possibilities of these structural elements.

And now when we come to examine the different classes of Radiates, we find that there are three that differ one from the other in exactly the manner in which a mathematician conceives that these elements may be combined with one another.

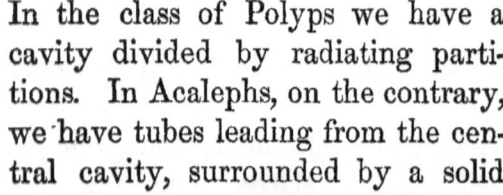

In the class of Polyps we have a cavity divided by radiating partitions. In Acalephs, on the contrary, we have tubes leading from the central cavity, surrounded by a solid gelatinous substance. In Echinoderms we have an outer solid wall, and these tubes transformed into independent organs, which wend their course in various ways in the interior, forming a compli-

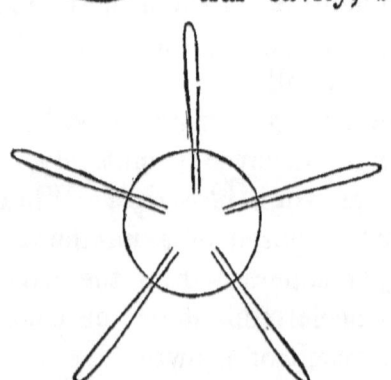

cated structure. So we have a plan in the construction of these animals similar to that which a mathematician would conceive. The mathematician to whom I appealed for the 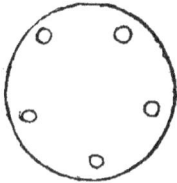 solution of the problem was entirely ignorant of natural history, and could not therefore have obtained his knowledge from the animal structures; and yet he at once devised these three as the only essential plans which could be framed upon the idea of a radiated structure around a vertical axis.

But now let us see what freedom and richness there is in the execution of that plan as represented in nature. In the first place, instead of a monotonous sphere in the living radiated animal we have a striking contrast between the two poles, one being essentially the base of attachment while the other is essentially a pole of expansion. The means of connection with the ground is composed of a different element from that by which the animal displays its activity. These two elements may be best shown in a profile view. The base of attachment presents a simple radiation; the sides of the spherical wedges are shown in the vertical lines upon the cylindrical body, while the upper end displays the radiation in the form of rays 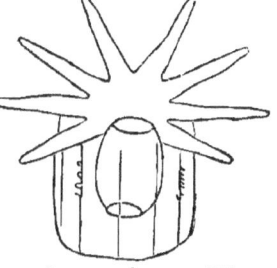 which communicate with the central opening. How much richer that is than the mere mathematical idea of a sphere! And yet you see how these elements are reducible to that very mathematical form.

We need only shorten these feelers and contract the upper part to reduce the whole upper region to a hemisphere. We need only lessen the height of the animal and reduce the base of attachment to a point, and we transform the whole body into a sphere.

Take now the Acaleph, or Jelly-fish. Here we have the same elements, only instead of preponderating cavities they are reduced to mere tubes, the intervals between them being composed of a solid mass. In addition to that we have elongated appendages or feelers. 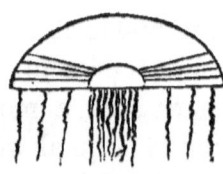 These can be better represented in a profile view. Around the margin of the bottom project a number of arms downwards. There is another set of appendages, which are only ornamental, introduced at the opening of the mouth. The marginal appendages may be more numerous than the rest, and there may be intermediate appendages; all these are similar parts, more or less numerous, with no new elements. Then again the diversity is increased by the tubes which radiate from the centre, instead of being straight, being variously complicated. 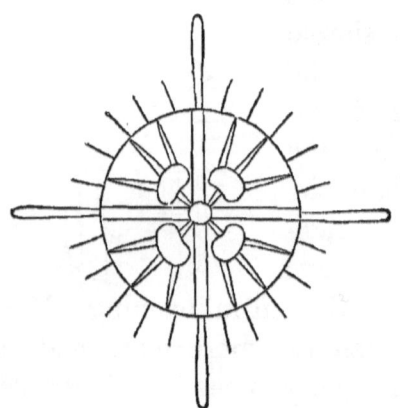 For instance, in the common white Jelly-fish, we have simple tubes radiating in four directions, alternating with other tubes which enlarge into a cavity, from which cavity proceed other small tubes radiating to the surface.

Let any one study the different varieties of Jellyfishes, and he will find in their forms all possible diversities of combination with the very few elements they contain. We have, for instance, long feelers and short ones between; then again long ones corresponding to the small tubes; and so on.

It is therefore evident that we have something similar to what may perhaps be better understood if I use the very familiar illustration of a musician who, taking a simple, familiar tune, plays an endless number of variations upon it, in each of which tne fundamental theme may be recognized. So this idea of radiation, with its simple elements, is played upon by the Almighty artist, and made into a multitude of living realities, to fill the world with variety. And the more we penetrate into the differences among these animals, the more do we see that between all there is an intellectual link which brings them into close relation and shows them to be but variations of an idea, and not the result of diverse circumstances and influences operating upon them. They were made what they are by an intellectual process which connects them all and combines them under one original plan. They are not the product of accident or of chance; and the evidence of the fact that they are the work of intellect may be derived from the facility with which our mind can grasp the idea which lies at the foundation of their structure, and generalize it. Let it once be understood that all radiate animals are built upon the plan of radiation, and let the elements of that radiation be discovered, and we at once perceive the intellectual thought that unites and combines them all.

But there are problems of still greater interest presented as we study these animals more in detail. You have been accustomed in these lectures to the presentation of no considerable details. I trust you will allow me this evening to enter into such details as will make it perfectly evident that when we analyze these structures we disclose the mental operations of the Creator at every step.

There is a third class of Radiates, more complicated in structure than either Polyps or Acalephs. It is the Echinoderms, including Star-fishes, Sea-urchins, and the like. I will select this class for illustrating the subject in the details.

In their external appearance they present three very marked modifications of form. Those known as 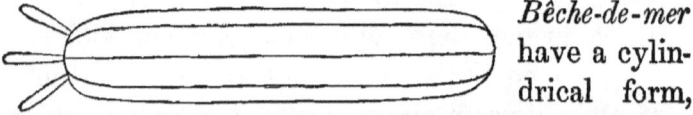 *Bêche-de-mer* have a cylindrical form, and in moving rest on one side. Their radiating organs are essentially developed at one end of the body, but continued by vertical rows which extend from one pole to the other. Naturalists call them Holothurians. Next to them we have those which 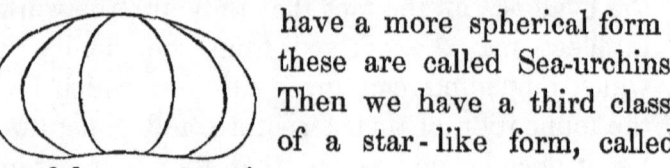 have a more spherical form; these are called Sea-urchins. Then we have a third class, of a star-like form, called Star-fishes, or Asterians.

Suppose the problem were presented thus: With the same structural elements to build a cylindrical, a spherical, and a star-shaped body. Ask an architect to build, with  the same number of pieces connected in the same way, a circular tower, an arched dome, and a pentagonal edifice! It will be a problem not easy to solve, especially if there is required the further condition that each structure shall be closed at the two ends.

When we examine the structure of these animals, the problem, instead of being so difficult as when presented in that form to the architect, becomes very simple. Indeed, the solution becomes at once apparent when we examine their structure, so much so, that, like the egg of Columbus, we are surprised that we did not know it before.

Let me, in entering into these details, give you all the representatives of this class, and then select the two extremes, namely, the spheroidal and the star-shaped, — the Sea-urchin and the Star-fish. It will be necessary to represent the Sea-urchin from three

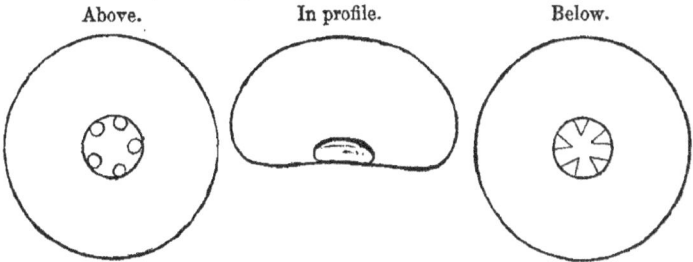

sides, above, in profile, and below. Below we have

an opening which is the mouth. It is armed with five jaws. Above we have a number of little plates, five of which are larger than the rest, each traversed by a hole. Then the interior is occupied by a number of very small plates. Now each one of these small plates on the upper side corresponds to an interval between the two jaws of the lower side. From the centre extend five rows to the periphery, both in the upper and lower part. These rows are built of a number of small plates, all of which are perforated with holes, and through these holes proceed tubes, by which the animal moves. The wider intervals between them are occupied by broader plates, and therefore fewer.

What are all these parts? The little plates at the summit are occupied by eyes; we have therefore five eyes. The broad plates which cover the surface of the Sea-urchin contain the soft parts, which come out through the rows, and by which the animal is enabled to move.

With these few very simple materials, how shall we build a Star-fish? Let us see how the Star-fish is constructed. Seen from above it presents a net-work of very minute plates, and there is no alternation of larger plates at all. They extend all over the surface. There is one point at which there is a large dot, and that is a sort of seam through which the water penetrates into the interior. There seems to be nothing like these parts in the Sea-urchin at first sight. But when we examine the Star-fish on the lower side, we find a very different structure. Here are five deep furrows.

Under those furrows project long tubes in every direction, being the organs with which the animal crawls about. They are evidently, so far as use is concerned, the same organs as those which project from the Sea-urchin. But in the Sea-urchin we have them all united, while in the Star-fish they are found only on the lower side.

Let us see what this furrow is made of, and how these tubes are connected with it. Each furrow is made up of a number of large pieces alternating with one another, and between them there are holes through which come the tubes by which the animal moves. Then, on the side of these broad tubes are smaller plates. Now these plates gradually taper in size until the whole is transformed into an angular furrow having the same structure all the way through. At the end of each ray in the Star-fish there is an eye. We have, therefore, everything on the lower surface in the Star-fish which we find over the whole body of the Sea-urchin, with the single exception of the small circle on the summit of the latter which is occupied by other plates.

If now we proceed to compare these arrangements, we cannot fail to see a certain analogy between them. Suppose I should split an orange into five parts and stretch those parts in every direction thus, forming a pentagonal figure. The sphere is transformed into a star. Now the point on the upper side of the Sea-urchin occupied by those small plates becomes divided and transferred, in the transformation to the Star-fish, to the five extremities of the rays; and the five eyes which I described in the Sea-urchin, occupying the apex, are now distributed to the five extremities of the rays of the Star-fish. So that, though at first sight they seem to be in a different position, they are in the same position. There is nothing changed except that the little circle, occupying the summit of the sphere, has been stretched, and its parts so multiplied, that it extends over the whole upper surface of the animal.

Now it becomes very apparent that this problem was easy; that it required only an increase of the elements in one direction to build up the Star-fish instead of the Sea-urchin.

Facts like this, I think, show the immediate working of mind in the construction of the animal kingdom. It is not a kind of work which is delegated to secondary agencies; it is not like that which is delegated to a law working its way uniformly; but is that kind of work which the engineer retains when he superintends and controls his machine while it is working. It is evidence of the existence of a Creator, constantly and thoughtfully working among the complicated structures that He has made.

In every class of animals we see the same thing. Let us take a familiar example, in order to show that what I have done with two animals, or two orders in the same class, or a great variety of members of the same class, may be done in the different parts of one and the same animal. I am prepared to demonstrate that such an animal as the Lobster, for instance, is built of the same elements from head to tail, only more or less modified according to the region they occupy; that the head, chest, claws, jaws, fins, and tail are all one and the same thing, only diversified according to the position and use of those parts.

As a starting-point let us examine the Worm. It is a cylindrical animal divided into a number of uniform joints, to which, among the higher families, are attached little paddles or oars for locomotion. In some there are a few bristles, perhaps three to each paddle, and they may be a little longer or shorter; but after all you perceive that every ring has its locomotive appendages, slightly modified according to the position of the body, but presenting great uniformity in all the parts, which are not bound together in two distinct regions.

It is, however, very different in the Lobster. There we have an anterior region which is very distinct from the posterior region. In the posterior region of the body the division into rings is at once very obvious. There are on the sides of the tail fin-like appendages, and under the tail other similar little appendages, which are all very different from those at the end of the tail. Then under the body are other appendages of greater length, with joints, and terminating in a little hook. There are five pairs of these, the front pair of which has a remarkable peculiarity, terminating in a large claw with two nippers, which may be moved one against the other with great force. Then in the head are several jaws which move one against the other; and in the extreme front two pairs of long feelers, the foremost of which contain a pair of eyes supported on a peduncle.

Let us examine the resemblance between these various parts. In the first place, whether one of these appendages has a larger or smaller number of joints is a matter of no consequence, because we find among Crustacea those in which the number of joints to a limb is greater or less. Whether the appendage terminates in a sharp claw, or whether the last joint is flattened, is also of no great consequence, because we find among Crustacea those in which one pair of legs is flattened in the shape of an oar, and others in which the end is shaped like a claw. And when we examine the different kinds of Shrimps which inhabit our coast, and compare them with one another, we find that there are those which, instead of flat oars, have, like the Crab, a slender branch at the side of the appendage; and still others which, instead of one oar, may have two. Therefore, between the leg and one of these oars under the tail of the Lobster there is only a difference in form, or in mode of execution; nothing different in their nature, because there is at the end of the tail a large pair of appendages which constitute a most powerful organ of locomotion, and that terminal pair differs from those under the tail only in size and power, not in structure. We are, therefore, as far as this, prepared to say that the large fins at the end of the tail, and the small fin-like appendages under the tail, and those slender ones under the main body and the legs, are all one and the same thing.

But here is the claw, which seems to be something entirely different in structure from the other appendages. Let us examine it more closely, and we shall see that it is not so different as it first appears. The leg is composed of a series of joints, three or four in number, and then a last joint tapering to a point. The last joint has a little hook at the side. We find the same thing in some Shrimps. Now let this hook be very much enlarged and you have the claw of the Lobster. And this last joint is made to play against the prong of the last joint but one, and makes a nipper. It is therefore identical in structure with all the other appendages; we have nothing new in the structure of the claw, only a modification of parts. Passing now to the jaws, we find little appendages armed with teeth, which, when brought one against the other, enable the animal to seize and crush its prey. These appendages are arranged on opposite sides and curved inwards, the inner margin being serried or variously armed. And what is peculiar to all Articulates, these appendages move horizontally towards one another, instead of perpendicularly. These jaws you perceive are nothing but legs made subservient to the use of seizing prey and crushing it and then pushing it into the alimentary canal. There are as many as six pairs of these jaws, one behind the other, all essentially of the same character as the claw, the leg, or the fin.

But what are these feelers? They come near the simplicity of structure of those appendages which we find in Worms, thread-like, and divided into a large number of close articulations. This organ is now made to subserve the purpose of touching or feeling. But it bears to the body the same identical relation as all the other parts.

If we examine how these appendages are connected with the rings of which the body is composed, we find them everywhere in the same position. The body is made up of rings from which arise lateral appendages in the shape of tubes; these tubes are jointed, and when highly complicated they assume the form of a claw, when flattened, the shape of an oar, when divided into a number of joints, feelers, and when made to move laterally, one set against another set, jaws. And the eye is attached to one of these feelers at the end. It is a feeler made extremely sensitive at its extremity, and especially sensitive to one kind of impression. But because it receives impression of light it is not therefore an eye like ours, having a close connection with the brain; it is only a footlike or claw-like organ placed on the summit of the appendage made sufficiently sensitive to receive an impression of light.

So, then, from one end of the body to the other, whatever be the function of these organs, we have only a different expression of one and the same thing.

And that this is the true view of the case we have evidence whenever we compare different animals of this class with one another. Take, for instance, the Crab and the Lobster. Take that singular animal called the Horseshoe crab which is found along our shore, and compare its rings with those of the Lobster. In the Lobster we have six pairs of jaws, six pairs of legs, six pairs of fins, a pair of eyes, a pair of feelers, and a broad terminal fin at the end of the tail. In the Horseshoe crab we find around the mouth six pairs of legs, no feelers, no jaws proper; but such is the complete demonstration of the identity of all these parts, that, while the tips of these legs are the feet with which the animal moves, the elbows of the same appendages are the surfaces which crush and bring the food to the mouth. The same appendage is, in other words, both a locomotive and a chewing organ, and they are placed in the same position as in the Lobster or Crab.

Nothing, therefore, could be plainer than that these various parts which are made subservient to such diverse purposes are essentially the same in structure, only differing in the execution. It is something akin to the device of man, to do as much work as possible with the smallest and simplest apparatus; and when the largest amount and greatest variety of work is produced by a particular invention, we consider the result as indicative of superiority of genius or inventive capacity. Here in the animal kingdom we see it illustrated to an extent which the best-trained mind can hardly follow, showing how far beyond our comprehension are the wonderful works of nature. Even though we can make ourselves conscious that they are built by mind, and that it has pleased the Maker of all things to give us a spark of that life which makes us to be His children, formed in His image, that evidence is nowhere stronger than in the fact that our mind is capable of studying those works to a limit which approaches to a comprehension of their wonderful relation to one another.

THE END.

www.ingramcontent.com/pod-product-compliance
Lightning Source LLC
Chambersburg PA
CBHW020234170426
43201CB00007B/418